CLASSIFY, EXCLUDE, POLICE

IJURR-SUSC Published Titles

Classify, Exclude, Police: Urban Lives in South Africa
and Nigeria
Laurent Fourchard

Housing in the Margins: Negotiating Urban Formalities
in Berlin's Allotment Gardens
Hanna Hilbrandt

The Politics of Incremental Progressivism: Governments,
Governances and Urban Policy Changes in São Paulo
Eduardo Cesar Leao Marques (ed.)

Youth Urban Worlds: Aesthetic Political Action in Montreal
Julie-Anne Boudreau and Joëlle Rondeau

Paradoxes of Segregation: Housing Systems, Welfare Regimes
and Ethnic Residential Change in Southern European Cities
Sonia Arbaci

Cities and Social Movements: Immigrant Rights Activism
in the US, France, and the Netherlands, 1970–2015
Walter Nicholls and Justus Uitermark

From World City to the World in One City:
Liverpool through Malay Lives
Tim Bunnell

Urban Land Rent: Singapore as a Property State
Anne Haila

Globalised Minds, Roots in the City: Urban
Upper-Middle Classes in Europe
Alberta Andreotti, Patrick Le Galès and
Francisco Javier Moreno–Fuentes

Confronting Suburbanization: Urban Decentralization
in Post-Socialist Central and Eastern Europe
Kiril Stanilov and Luděk Sýkora (eds.)

Cities in Relations: Trajectories of Urban
Development in Hanoi and Ouagadougou
Ola Söderström

Contesting the Indian City: Global Visions and the Politics
of the Local
Gavin Shatkin (ed.)

Iron Curtains: Gates, Suburbs and Privatization
of Space in the Post-socialist City
Sonia A. Hirt

Subprime Cities: The Political Economy of Mortgage
Markets
Manuel B. Aalbers (ed.)

Locating Neoliberalism in East Asia: Neoliberalizing
Spaces in Developmental States
Bae-Gyoon Park, Richard Child Hill and Asato Saito (eds.)

The Creative Capital of Cities: Interactive Knowledge of
Creation and the Urbanization Economics of Innovation
Stefan Krätke

Worlding Cities: Asian Experiments and the Art of Being Global
Ananya Roy and Aihwa Ong (eds.)

Place, Exclusion and Mortgage Markets
Manuel B. Aalbers

Working Bodies: Interactive Service Employment and
Workplace Identities
Linda McDowell

Networked Disease: Emerging Infections in the Global City
S.Harris Ali and Roger Keil (eds.)

Eurostars and Eurocities: Free Movement and Mobility in
an Integrating Europe
Adrian Favell

Urban China in Transition
John R. Logan (ed.)

Getting into Local Power: The Politics of Ethnic
Minorities in British and French Cities
Romain Garbaye

Cities of Europe
Yuri Kazepov (ed.)

Cities, War, and Terrorism
Stephen Graham (ed.)

Cities and Visitors: Regulating Tourists, Markets, and
City Space
Lily M. Hoffman, Susan S. Fainstein,
and Dennis R. Judd (eds.)

Understanding the City: Contemporary and Future Perspectives
John Eade and Christopher Mele (eds.)

The New Chinese City: Globalization and Market Reform
John R. Logan (ed.)

Cinema and the City: Film and Urban Societies in a
Global Context
Mark Shiel and Tony Fitzmaurice (eds.)

The Social Control of Cities? A Comparative Perspective
Sophie Body-Gendrot

Globalizing Cities: A New Spatial Order?
Peter Marcuse and Ronald van Kempen (eds.)

Contemporary Urban Japan: A Sociology of Consumption
John Clammer

Capital Culture: Gender at Work in the City
Linda McDowell

Cities After Socialism: Urban and Regional Change and
Conflict in Post-Socialist Societies
Gregory Andrusz, Michael Harloe and Ivan Szelenyi (eds.)

The People's Home? Social Rented Housing in Europe
and America
Michael Harloe

Post-Fordism
Ash Amin (ed.)

Free Markets and Food Riots
John Walton and David Seddon

CLASSIFY, EXCLUDE, POLICE

Urban Lives in South Africa and Nigeria

LAURENT FOURCHARD

Registered Office(s)
John Wiley & Sons, Inc., 111 River Street, Hoboken, NJ 07030, USA
John Wiley & Sons Ltd, The Atrium, Southern Gate, Chichester, West Sussex, PO19 8SQ, UK

Editorial Office
9600 Garsington Road, Oxford, OX4 2DQ, UK

For details of our global editorial offices, customer services, and more information about Wiley products visit us at www.wiley.com.

Wiley also publishes its books in a variety of electronic formats and by print-on-demand. Some content that appears in standard print versions of this book may not be available in other formats.

Library of Congress Cataloging-in-Publication Data

Name: Fourchard, Laurent, author.
Title: Classify, exclude, police : urban lives in South Africa and Nigeria
    / Laurent Fourchard.
Other titles: Trier, exclure et policer. English
Description: Hoboken, NJ : John Wiley & Sons, 2021. | Series: IJURR studies
    in urban and social change book series | "Original French edition,
    Trier, exclure, policer. Vies urbaines en Afrique du Sud et au Nigeria ©
    2018 Presses de la Fondation nationale des sciences politiques"–Verso.
    | Includes bibliographical references and index.
Identifiers: LCCN 2020043244 (print) | LCCN 2020043245 (ebook) | ISBN
    9781119582625 (cloth) | ISBN 9781119582649 (paperback) | ISBN
    9781119582670 (adobe pdf) | ISBN 9781119582656 (epub)
Subjects: LCSH: Marginality, Social–South Africa. | Marginality,
    Social–Nigera. | Discrimination–South Africa. |
    Discrimination–Nigeria. | Urban policy–South Africa. | Urban
    policy–Nigeria. | Police–South Africa. | Police–Nigeria. | South
    Africa–Ethnic relations. | Nigeria–Ethnic relations.
Classification: LCC HN801.Z9 M264613 2021 (print) | LCC HN801.Z9 (ebook)
    | DDC 305.5/6809669–dc23
LC record available at https://lccn.loc.gov/2020043244
LC ebook record available at https://lccn.loc.gov/2020043245

Set in 11/13pt Adobe Garamond by SPi Global, Pondicherry, India

10 9 8 7 6 5 4 3 2 1

Contents

Series Editors' Preface

IJURR Studies in Urban and Social Change Book Series

The IJURR Studies in Urban and Social Change Book Series shares IJURR's commitments to critical, global and politically relevant analyses of our urban worlds. Books in this series bring forward innovative theoretical approaches and present rigorous empirical work, deepening understandings of urbanisation processes, but also advancing critical insights in support of political action and change. The Book Series Editors appreciate the theoretically eclectic nature of the field of urban studies. It is a strength that we embrace and encourage. The editors are particularly interested in the following issues:

- Comparative urbanism
- Diversity, difference and neighbourhood change
- Environmental sustainability
- Financialisation and gentrification
- Governance and politics
- International migration
- Inequalities
- Urban and environmental movements

The series is explicitly interdisciplinary; the editors judge books by their contribution to the field of critical urban studies rather than according to disciplinary origin. We are committed to publishing studies with themes and formats that reflect the many different voices and practices in the field of urban studies. Proposals may be submitted to editor in chief, Walter Nicholls (wnicholl@uci.edu), and further information about the series can be found at www.ijurr.org.

Walter Nicholls
Manuel Aalbers
Talja Blokland
Dorothee Brantz
Patrick Le Galès
Jenny Robinson

Acknowledgements

This book is located at the crossroads of comparative urban studies, history and political science using historical and ethnographic methods to provide multiple historicity, from the colonial to contemporary periods. It is largely the product of a trajectory related to the functions I have been able to perform, to the modalities of the research conducted in South Africa and Nigeria but also and above all to professional and friendly encounters that have nourished my personal reflection for the past twenty years.

As director of the French Institute for Research in Africa (IFRA Nigeria) between 2000 and 2003, I benefitted from the support of researchers who introduced me to the contemporary history of Nigeria including Olufunke Adeboye, Olutayo Adesina, Saheed Aderinto, Rufus Akinyele, Isaac Olawale Albert, Toyin Falola, Dele and Peju Layiwola and Rasheed Olaniyi. I have also benefitted from the support of IFRA during all these years to present my work during seminars or master classes in collaboration with Brian Larkin, Élodie Apard, Émilie Guitard and Ismael Maazaz. Many friends and colleagues sensitise me into topics of urban governance, security, party politics and social history in the first years of 2000 (Claire Bénit-Gbaffou, Catherine Coquery-Vidrovitch, Alain Dubresson, Philippe Gervais-Lambony, Odile Goerg, Sylvy Jaglin, Alan Mabin, Dominique Malaquais and AbdouMaliq Simone).

Appointed as a research fellow in 2004 at the Centre for African Studies (CEAN renamed LAM in 2011) of the Institute for Political Studies (IEP) in Bordeaux, I have strengthened my interest in comparative research between South Africa and Nigeria through numerous seminars and informal discussions with many Bordeaux-based scholars: Jean-Nicolas Bach, Louise Barre, Léa Barreau-Tran, Jean-Philippe Berrou, Vincent Bonnecase, Chloé Buire, Michel Cahen, Hélène Charton, Denis Constant Martin, Jean-Pierre Chrétien, Mathias Delori, Christine Deslaurier, Dominique Darbon, Alain Durand-Lasserve, Vincent Foucher, Didier Galibert, Marc-Eric Gruénais, Comi Toulabor, Daouda Gary-Tounkara, Jean-Hervé Jézéquel, Alessandro Jedlowski, Chloé Josse Durand, Jean-Baptiste Lanne, Claire Médard, Hervé Maupeu, René Otayek, Wyclife Othiso, Gilles Pinson, Céline Thiriot, Lucie Revilla, Alain Ricard, Ophélie Rillon, Emmanuelle Spiesse, Magali della Sudda and Djemila Zeneidi.

As a visiting scholar at the University of Cape Town (UCT) in 2008–2009, several colleagues guided me in my fieldwork and my readings and many invited me to present the results of my research: Simon Bekker, Clive

Glaser, Maanda Mulaudzi, Sophie Oldfield, Susann Parnell, Edgar Pieterse, Jeremy Seekings, Christopher Saunders, Jennifer Robinson, Elrena van der Spuy, Clifford Shearing, Steven Robins and Kees van der Waal. At the same time, I and Simon Bekker of the University of Stellenbosch carried out a broad collective research on the government of Africa's cities, which greatly stimulated me in my comparative approach. At the same time, I shared many of my research interests with French and South African PhD students and researchers: Claire Bénit-Gbaffou, Julie Berg, Chloé Buire, Lydie Cabane, Adrien Delmas, Vincent Darracq, Sophie Didier, Jeremy Grest, Liela Groenewald, Mariane Morange, Elizabeth Peyroux, Aurelia Segatti, Myriam Houssay Holzschuch, Marie Huchzermeyer and Jean-Fabien Steck. In 2011, a stay at Oxford under the Oxpo program allowed me to receive in-depth feedback on my work on vigilantism and the manufacture of delinquency in Nigeria and South Africa by Jocelyn Alexander, Elizabeth Cooper, Adam Higazi, Gary Kynoch, Jimam Lar, Kate Meagher, David Pratten and Olly Owen.

Joining the Centre for International Relations (CERI) in 2015 reinforced my interest for a historical and economic sociology of politics largely inspired by the European network of the analysis of political societies (REASOPO) coordinated by Jean-François Bayart and Béatrice Hibou and by co-organising the seminar on 'violence and citizenship in Africa' with Richard Banégas, Roland Marchal and Sandrine Perrot or by participating in the project on 'the social and political life of identity papers' coordinated by Richard Banégas and Séverine Awenengo Dalberto. Many chapters bear the traces of these discussions. My previous work on police work found a comparative extension outside the African continent by joining a research group on comparative vigilantism (GRAV) set up by Laurent Gayer and Gilles Favarel-Garrigues. I also developed an interest in cross-cutting research outside Africa, through my participation at the Urban School of Sciences Po, with a team including Fatoumata Diallo, Brigitte Fouilland, Charlotte Halpern, Patrick Le Galès, Côme Salvaire, Dennis Rogers and Tommaso Vitale. The feedback of master students of the urban school was also central to me.

This book has also been fuelled by numerous discussions with the members of the editorial boards of *Africa*, *the Journal of African History* and of the *International Journal of Urban and Regional Research* (*IJURR*) (Rifke Jaffe, Jennifer Robinson, Lisa Weinstein among many others) and members of the *Politique Africaine*, including Marie-Emmanuelle Pommerolle, with whom I was co-editor-in-chief from 2009 to 2012. I benefitted from intense exchanges conducted in the summer schools in comparative urban studies of the journal *IJURR*, thanks to the complicity of Claire Colomb, Yuri Kazepov, Patrick Le Galès, Eduardo Marques, Jeremy Seekings and Jennifer Robinson as well as those conducted in Marrakesh Winter School on the temporality of politics organised by the Chair of Comparative African Studies at the University of Rabat (Jean-François Bayart) and by a team of researchers from the Association of the journal *Politique Africaine* (Séverine Awenengo Dalberto, Fred Eboko, Thomas Fouquet, Nadia Hachimi, Didier Péclard, Marieme N'Diaye, Boris Samuel, Etienne Smith).

Jean-François Bayart, Frederick Cooper, Peter Geschiere, Fabien Jobard, Patrick Le Galès and Ibrahima Thioub brought me central elements when defending my *Habilitation* in Sciences Po Paris in 2015, which helped me to enrich my original manuscript. The French version of the book would not have been possible without their advice. I would also like to thank Emmanuel Blanchard, Vincent Bonnecase, Daouda Gary-Tounkara and Joel Glasman for reading some chapters and for their bibliographical suggestions. Presentation of the French version of the book has received many critical comments from Elodie Apard, Jean-Pierre Bat, François Bonnet, Chloé Buire, Corentin Cohen, Julie Clerc, Gilles Favarel-Garrigues, Laurent Gayer, Paul Grassin, Béatrice Hibou, Patrick Le Galès, Virginie Malochet, Claire Médard, Thierry Oblet, Daniel Sabbagh, Marianne Saddier and Djemila Zeneidi whose suggestions helped me improve the English version. Jennifer Robinson who encouraged me for years to write a book, provided a considerable number of suggestions for revising the English version. I would like to warmly thank her as well as Walter Nicholls, the chief editor of the SUSC series and Jacqueline Scott for being very supportive. This English version was made possible, thanks to the executive director of the Presses de Sciences Po, Julie Gazier, thanks to the financial support of CERI and the Urban School of Sciences Po and to the translator Susan Taponier.

I would like to warmly thank Joseph Ayodokun (in Ibadan), Gboyega Adebayo and Joseph Akinniyi (in Lagos) and David Agige (in Jos) who helped me interview civil servants, politicians and union officials in the neighbourhoods mentioned in the book. I also thank them for the translations made in Yoruba (Lagos and Ibadan) in Hausa, Afizere and Anaguta (in Jos). Their work has considerable influence and I wish them success in achieving their doctoral thesis.

Emmanuelle Spiesse, Marcel, Gaspard and Achille Fourchard have been infinitely patient and unwaveringly supported me throughout the years. This book is dedicated to them.

Classify, Exclude, Police

Nigeria, 2006: Human Rights Watch published a report on discrimination against the country's non-indigene*[1] populations:
'The population of every state and local government in Nigeria is officially divided into two categories of citizens: those who are indigenes and those who are not. The indigenes of a place are those who can trace their ethnic and genealogical roots back to a community of people who originally settled there. Everyone else, no matter how long they or their families have lived in the place they call home, is and always will be a non-indigene.' (Human Rights Watch, April 2006, p. 1.)

The report indicates that many states refuse to employ non-indigenes in the civil service, discriminate against them in the provision of basic services, and often deny them the right to stand for office in local and state government elections, thereby treating them as second-class citizens. Furthermore, the report asserts that the division between indigenes and non-indigenes has led to extreme violence in some localities: 1,000 people died in the city of Jos (in the centre of the country) in September 2001, more than 600 in the small town of Yelwa (200 km from Jos) during the first half of 2004, and several hundred in 1997 and 2003 in the city of Warri (Niger Delta).

South Africa, May 2008: xenophobic violence engulfed the whole country during the month of May, leaving 60 dead, 700 injured, and more than 100,000 displaced. A third of the victims were South Africans, although foreigners from other African countries were the main targets (Landau 2011, p. 1). The violence began in the township of Alexandra in Johannesburg, then spread to other townships chiefly in the province of Gauteng, and later to the cities of Cape Town and Durban. The 140 zones involved were mostly townships and informal urban areas.

The violence of the attacks was unspeakable. For the moment, we would simply note that it was grounded in the exclusion of a group based solely on nationality (other than South African) or origin (non-indigene), and that it took place on a national scale and in urban environments.[2] National affiliation or supposed origin are only one among numerous repertoires of exclusion and one of the categories that potentially generates the use of violence, but their repetition and widespread protean nature – offences against the integrity of

Classify, Exclude, Police: Urban Lives in South Africa and Nigeria, English Language First Edition. Laurent Fourchard.
© 2021 John Wiley & Sons Ltd. Published 2021 by John Wiley & Sons Ltd.

persons, mob violence, repression by security forces – and the countless forms of exclusion are indeed at the core of the historiography of these two countries.

Metropolises function as command posts, overseeing a concentration of population, production and consumption. As such, they offer an ideal observation point for studying the day-to-day practices of power and the genealogy of forms of exclusion. Lagos, Ibadan and Kano in Nigeria and Cape Town and Johannesburg in South Africa have been metropolises for over a century (see Figures I.1 and I.2). All of them house government agencies and influential political networks.[3] By the end of the nineteenth century, they had become leading labour markets at the regional or countrywide level, and their rapid growth (see Table I.1) soon gave rise to new forms of poverty and social violence (unemployment, delinquency, maltreatment, prostitution, gangsterism, procuring) and problems integrating migrant populations. Their increasing social diversity generated a profusion of discourses and they became privileged

FIGURE I.1 States and cities in Nigeria **Source:** Realised by Christine Deslaurier. IRD, UR 102, 2007.

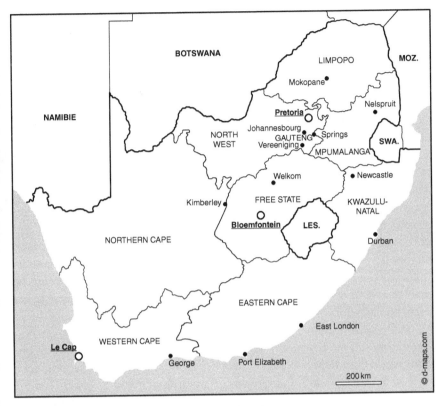

FIGURE I.2 Provinces and cities in South Africa.

TABLE I.1 The population of Lagos, Ibadan, Kano, Johannesburg and Cape Town (in thousands)[5]

	1866	1891	1911	1952	1963	1970	1991	2010
Lagos	25	32	73	272	542	1266	5195	8048
Ibadan	100	120	175	459	427	998	1835	2551
Kano	30			127	255	882	2167	2826
	1866	1891	1904	1951	1960	1970	1996	2011
Johannesburg	—	3	327	969	1247	1561	2638	4434
Cape Town	22	51	77	500	803	1300	2565	3740

places for producing knowledge and testing, developing and implementing new apparatuses of power.[4] These apparatuses contributed significantly to turning the metropolises into laboratories for exclusion and the use of violence. Some initiatives were introduced by state agents to target categories of people whose socialisation to urban life was deemed problematic (temporary migrants, non-natives*, delinquents, children in need of care, single women). Over time such instruments became permanent features of city life, but they remained politicised, conveying values that embodied a particular interpretation of society and ideas about how to regulate it (Lascoumes and Le Galès 2004, p. 13).

From the early twentieth century onwards, South African labour policies divided workers into two separate groups: on one hand, a category of urban workers who were to be made into stable residents by granting them rights related to housing, employment and family life, and on the other, a population of temporary migrants destined to return to the countryside once their labour contracts were completed. For the members of this second group, the hostel* – or more precisely the assignment of a bed ('bedhold') – became the institution structuring their daily lives, as well as relationships with their employers and administrative authorities, fellow workers at the hostel, apparently favoured neighbours living nearby in family quarters and women whose unauthorised residence depended on the goodwill of the men to whom they had to be attached (Ramphele 1993). During the same period, labour policies in Nigeria led to the creation of a new category of urban resident called 'non-native' – defined in opposition to 'native' – which was the norm at the time. These policies authorised the presence of migrants needed by the colonial economy, but required them to reside in reserved neighbourhoods and placed them under a separate authority to avoid diminishing the power of native chiefs.

The historical invention of these categories is at the heart of the processes of exclusion and the reification of differences between natives and non-natives, and between urbans and temporary migrants, which had enduring legacy in post-colonial and post-apartheid periods. In the 1930s, further categories were added in both countries to define and classify urban youth as delinquents, children in need of care and minor girls in need of protection. When South African social workers and their British imperial counterparts in Nigeria embarked on a mission to have boys released from prison and protect girls from the dangers of street life, they set up social services that criminalised the presence of these young people on the streets. In Lagos, they sent the boys to the countryside, and prohibited minor girls from street trading. Girls were regularly rounded up, forced to undergo gynaecological examinations, and confined to hotels in the company of prostitutes, thereby arousing the indignation of their parents. The notion of delinquency as a form of criminal behaviour came into being during this period, but it is difficult to ascertain whether it applied to street children, girl street vendors, occasional thieves or hardened criminals. A similar ambiguity surrounds contemporary local expressions (*tsotsis* or *skollies* in South Africa, *boma boys* or *jaguda boys* in Nigeria) used to describe groups of boys engaged in activities on the borderline between legality and illegality, ranging from shoplifters and groups of neighbourhood mates to hierarchically organised gangs.

These examples attest to the new mode of governing populations introduced during the colonial period. Migrants, non-natives, delinquents, children in need of care, minor girls and single women became administrative categories that had little in common other than being part of a nomenclature designed to rule by classification – a process that paralleled the invention and reification of ethnic groups in the countryside.[6] Indeed, the new categories constituted more than just an administrative taxonomy: they were also associated with rights (regarding work, access to housing or place of residence), punishments (prison, deportation, fines, flogging) and prohibition (from circulating freely, engaging in trade, working, living alone or with a family). The individuals concerned used these categories to define and describe themselves as well as to describe, stigmatise and exclude other groups they considered their opponents. As a result, these categories left a deep imprint on the collective imagination long after such social engineering was abandoned.

In the 1930s, new security apparatuses were also developed by non-state actors at the neighbourhood level. After identifying the most obvious threats, local organisations in South Africa and Nigeria introduced schemes for policing everyday life in low-income districts. These actors were given substantial power for the reason that the authorities had neither the resources nor a compelling need to ensure a police presence in areas that played a minimal role in the colonial economy. Such organisations acquired considerable operational autonomy and, in some respects, actually governed the neighbourhood, especially at night. They were free to use undue violence against unruly youths or 'foreigners' whose access to the neighbourhood was regulated after certain hours. By drawing the boundaries between insiders to be integrated and outsiders, they built a political community at the neighbourhood level.

These organisations or their successors still operate in the two countries today, but their *modus operandi* has been partly transformed. The violence is perhaps better regulated nowadays; corporal punishment no longer enjoys the same legitimacy, and it has become more discreet or rare but it has not disappeared. Neighbourhood policing organisations have become more bureaucratised, politicised and, in some cases, feminised. They now charge low rates for their services, rather than performing them for free. But, as in the colonial period, policing still consists in identifying specific threats to neighbourhood cohesion and controlling target populations.

Finally, other apparatuses are used at the micro level of bureaucratic and political spaces (local government offices, the residence of a political boss) and public spaces (the street, bus stations, markets). Access to these places is constantly being challenged and brokered between individuals in positions of authority (civil servants, trade union leaders, godfathers, governors) and subordinate actors (street vendors, bus drivers, unemployed workers, students) seeking a service, an authorisation, a document, a stall in a market or a place at university. Observing the myriad negotiations between street-level bureaucrats, private agents working on behalf of the state and economic players whose livelihoods depend on the street provides an up-to-date picture of the opportunities for inclusion in a clientelistic network or in a local political community as well as the forms of discrimination at play in granting access to a service, a job or a space for trading. Above all, these negotiations reveal a whole range of diverse practices. For a street vendor, the process may involve bargaining for a reduction in the amount to be paid to a municipal tax collector; for a student, complying with what local government agents present as the rules for obtaining a certificate and finding a job or a place at university; for a tax collector at a bus station, being paid by the road transport union to intimidate or attack bus drivers who refuse to yield to union authority; for an unemployed unskilled worker, to benefit from the largess of a godfather, including food, in exchange for intimidating his political opponents if need be during electoral periods.

By exploring these varied apparatuses, we can measure the effects of classifying populations in terms of exclusion and inclusion, the violence they sometimes engendered, and the forms of social differentiation they brought about. For example, one might examine how state agents identified and analysed different (or similar) problems in Kano, Lagos, Cape Town and Johannesburg and how this process led to policies that simultaneously altered the limits of the state, claimed to govern conduct and produced social differentiation; or we might look at why the relationships between political bosses and their clients in Ibadan and Lagos, which for many years were quite similar, began moving in different directions in recent times, conditioning the violent (or non-violent) nature of mobilisations, state intervention and the integration of city dwellers in political networks; or we might question why the everyday work of patrolling neighbourhoods, which appeared to be identical in Ibadan and Cape Town, has had radically different effects on the construction of community boundaries within these cities, on the use of coercion, and on what these practices tell us about the nature of a state that outsources its security functions in this way.

This book does not tell the story of how most of the population was denied political rights – the foundation of apartheid and colonial and military regimes – nor of the inequalities and the enrichment of a racial or political minority at the expense of the majority, or of repression and attempts by the police or the army to subject citizens. Instead, bearing in mind this historical matrix, it invites us to step away from the national frameworks to study the myriad of urban arrangements used to manufacture exclusion. By articulating insights across the various local, national, imperial and global levels, and the colonial, apartheid and contemporary periods, it will bring to the fore the everyday practices of power and a genealogy of different forms of classification, exclusion and policing. By focusing on power relationships in daily life and understanding objects that extend across the twentieth century into the twenty-first, I am seeking to open up a dialogue including equal parts of history, political sociology and comparative urban studies. I am seeking to account for the chaotic historical formation of the state – which could not be reduced to its fragile or weak dimensions – and an unforeseeable changing urban social reality that cannot be adequately explained by planning nor by informality. Essential components of the repertory of public actions such as providing security, attempting to monopolising violence, and producing or reproducing social, racial or generational differences are not restricted to state agents but carried out by numerous social groups that change their concrete implementation, often in very accidental ways.

This process echoes the genealogical analysis of Foucault, whose aim is to retrace the historical conditions of emergence of specific apparatuses of power. Foucault was largely influenced by Nietzsche's own interpretation of history, as shown in his key text published in 1971 (Foucault 1971, pp. 1004–1024). Genealogy aims to 'detect the singularity of events outside of any monotonous finality' (Foucault 1971, p. 1004). It requires the meticulousness of knowledge, a large number of materials piled up, details, random beginnings, accidents, small deviations or complete reversals, errors, misjudgements, miscalculations. Foucault is said to have challenged 'traditional history', but according to his friend, the historian and philosopher Paul Veyne, Foucault believed only in the truth of the facts, of the countless historical facts that fill all the pages of his books; he was sceptical towards any universal concepts and started from the concrete practices of the state, and ordinary places of power rather than from general and well-known ideas (Veyne 2008, p. 9, 19, 33). Interpreting history as a series of accidents and considering 'that the forces at stake in history do not obey a destination or a mechanism but the randomness of the struggle' (Foucault 1971, p. 1016) was quite unusual in the 1970s. Foucault's genealogy was opposed to meta-historical forms of writing (Foucault 1971, p. 1005), to historians who focus excessively on causal relationships (Veyne 2008, pp. 38–39), and to a history that does not take into account ordinary mechanisms of power (Foucault 1971, p. 1019). As David Garland explains (2014, p. 372), 'genealogical analysis traces how contemporary practices and institutions emerged out of specific struggles, conflicts, alliances, and exercises of power, many of which are nowadays forgotten'. This book is an attempt to discover exercises of power, often located outside state power, sometimes within it.

In keeping with sociological traditions from its founding fathers to Michel Foucault and Paul Veyne, this book calls into question the antinomy between past and present. In an age of fragmented, hyper-specialised and standardised knowledge, this choice reflects a scientific conviction – a unified conception of the social sciences including history – and a taste for intellectual and methodological cross-pollination and comparative approaches. Such an endeavour entails articulating periods, and in this case, taking stock of the legacy from the colonial era and earlier periods, some of which are present in the forms of postcolonial power (Bayart and Bertrand 2006, p. 142). It means searching for the traces of the past in the present day (including the present of the historical actors) by making intelligible how 'the things of the past are objectivised and crystallised in mental structures, in material things, in habitus' (Noiriel 2010). This approach is characteristic of comparative historical sociology, which calls into question the opposition between idiographic and nomothetic disciplines (Bayart 1989; Déloye 1997) and of sociohistory, which emphasises the need to historicise social relationships viewed as products of power relationships that have become solidified, objectified, and naturalised over time (Noiriel 2012, 2010).

This dialogue between history and social sciences are central in the tradition of interdisciplinary knowledge in African studies (Balandier 1985 [1955]; Coquery-Vidrovitch 1993; Ferguson 1999; Freund 2007; Mbembe and Nuttall 2004) which in some cases have entered into a productive dialogue with urban studies of the continent (Myers 2011; Simone and Pieterse 2017; Robinson 2006; Simone 2004). These have helped to include a more diverse historical experience in understanding the urban world and possible peculiarities of urban histories of the global south. Despite this tradition, a very specific use of history needs to be interrogated within some segments of urban studies. First, the historical experience is rarely based on a dialogue with the long tradition of African history. Second, while ethnographic methods have become central in so many disciplines, historical methods, i.e. the patient time-consuming archival work and collection of oral and material sources, often seems to be marginal, leaving urban scholars with a second-hand approach and a relative neglect of historical everyday life. Third, many scholars have not totally avoided a form of presentism. Presentism – or the rise of the category of the present until this comes a ubiquitous evidence – partly consists of reinterpreting the past according to this present (Hartog 2003, p. 223). A genealogical analysis is suggested as a possibility to avoid that risk while questioning the often-marginal place devoted to the historicised dimension of urban life.

Practically, this means that there is a need to take contemporary and colonial periods seriously, i.e. systematically giving priority to primary sources with the same density of information. It has been of paramount importance to constitute a body of historical and ethnographical sources around shared questions. The historical sources include colonial, national and local archives consulted since 2000 at various sites in Nigeria (Ibadan, Kaduna, Enugu), South Africa (Pretoria, Cape Town, Johannesburg) and Great Britain (London, Oxford), and the local and national press archives in both countries, along with several dozen interviews with 'elders' on certain topics. The ethnographic materials come from a familiarisation with the environment of the neighbourhoods investigated in Cape Town, Ibadan and Lagos, repeated interviews with dozens of police organisation members, civil servants, union members and traders, together with participatory observation whenever possible (volunteer patrols in Cape Town, local governments in Ibadan and Lagos).[7] Rancière (1987) invites us to move between a word divided 'between those who explain and those who listen'. As a French academic living and working in Nigeria and South Africa I was first and foremost educated by the many informants I had the chance to meet. Cross-checking historical sources with ethnographical materials generated dialogue between literatures that usually do not communicate their findings and brought to light particular historical configurations that raised questions about changes in national policy (colonial/post-colonial, military/post-military, apartheid/post-apartheid), which are often taken as the narrative starting point for want of longitudinal studies.

Many scholars have recommended provincialising Western urban analysis to achieve a more comprehensive view of global urbanisation, to give priority to a post-colonial approach, and to call for a localisation of theoretical production (Robinson 2012; Roy 2011). If this book explores colonial and postcolonial history to figure out a peculiar historicity that may not fit a ready-made definition of the urban, my aim is not to take part in developing a 'Southern theory' that would abstract from, circumvent or challenge 'Western thought'. I prefer to read academics promoting Southern theory as cultural intermediaries who have raised these issues precisely because they have knowledge in different fields and different academic institutions (northern and southern). James Ferguson reminds us that 'Thinking does not mean thinking from a single point of view, but thinking from more than one point of view at the same time' (Ferguson 2012), a comment made in reference to Jean and Jane Comaroff who, like many 'intellectuals of the South', have positioned themselves in the space between them (Comaroff and Comaroff 2012).

I have adopted a similar approach in exploring this space between different academic traditions. For historical reasons, African universities are too often in a subaltern position in the production of knowledge that has received worldwide audience in the social sciences. A first step towards rebalancing this unequal production is to use the considerable and often neglected research produced in African universities. There is a long tradition of research in South African and Nigerian universities not always exposed in urban studies journals but which constitute the fundamental basis of this book. To produce new knowledge and new facts on urban Africa is another necessary step, not a technical point but a need for an exhaustive empirical research for which not abundant available data pre-existed for a number of issues (Pieterse and Parnell 2016). This is also an epistemological choice situated in a grounded theory that provides theoretical insights based on empirical data (Glaser and Strauss 1967) that are necessarily provisional and revisable (Robinson and Roy 2016) and only possible through a constant process of exploring the tiny minutiae of urban life (Simone and Pieterse 2017).

I have often been asked what is the point in drawing parallels between countries as different as South Africa and Nigeria, or cities like Cape Town and Ibadan that have nothing in common. This work is part of this renewal of the comparison in urban studies. Jennifer Robinson invites us to think of a global comparative urbanism and to shift the research produced within the Western European world and the United States to take into account different urban experiences in the South and to produce South/South and South/North comparisons (Robinson 2016; Robinson and Roy 2016). She calls for thinking together scales, flows and peculiar histories and to compare iconic cities with more ordinary ones and thus to move out of any exceptional visions of cities (Robinson 2006). In the African continent especially, there is a need to rethink hasty classifications. African cities are often qualified as 'colonial', 'post-colonial, 'informal, 'in crisis' and 'neoliberal' as if they were only laboratories or testing grounds for broader international dynamics over which local or national actors seem to have little influence (Fourchard 2011a). Comparative methods also limit the use of superlatives and distance ourselves from national debates (McFarlane 2010; Simone 2010). They are a way to move outside methodological nationalism, which, until recently, was common in urban sociology that carries the risk of reifying national experiences (Le Galès 2019, p. 34). South Africa and Nigeria are no exception, their historiographies are often nationally oriented, and have sometimes lent credence to the idea of incommensurable trajectories. The wish to move beyond an assumption of incommensurability follows some previous collective efforts to compare politics and policies in different cities of the continent (Bekker and Fourchard 2013). Comparing simultaneously avoids the pitfall of a 'new localism' commonly found in urban studies, which investigates local strategies, the capabilities of local actors, and local regulations, while overlooking more comprehensive national, regional or global transformations (Le Galès 2003, p. 27). There is a long tradition in sociology, geography and history informing the particularity of only one metropolis without using comparative tool; this work is also sensitive to a recent call for using more history and political sociology in comparative urban studies (Le Galès 2019).

Comparison is used here as a way to identify singularities: it means highlighting differences rather than dissolving them, sharing questions instead of answers. I do not intend to 'compare one to the other' but rather 'one *and* the other', to foreground particularities and contingencies rather than misleading likenesses (Bayart 2008a). It does not emphasise connections or flows between cities of the continent or elsewhere nor does it try to find variables of relatively similar cases. It wishes instead to allow each case to inform assessments of the other while pointing out particularities through a thorough description of practices and discourses in historically situated contexts. A creative attempt of post-colonial approach opts for a view on fragmented urban lifestyles and worlding experiences that give preference to unstable practices, fluidity, informal relations and transnational connections (De Boeck and Plissart 2004; Fouquet 2014; Malaquais 2006; Nuttall and Mbembe 2005; Roy 2016; Simone 2001). While sharing a rejection of any linear or teleological vision of history I think that this ethnography of fragments does not always help to explore what Foucault calls historical discontinuities and events (Foucault 1971, pp. 1015–1017, 1972, p. 1141). 'Event means not a decision, a treaty, a reign or a battle but a balance of forces that is reversed, a power confiscated, a vocabulary taken up and turned against its users, a domination that weakens, relaxes, poisons itself, another that makes its entrance, masked' (Foucault 1971, p. 1016).[8] How might the change of power relationships in its tiny details reflect specific configurations of institutions, transgression of established discourses and naturalised practices?

In other words, comparing a genealogy of classification, exclusion and police will bring to the fore the everyday practices of power in past and present urban lives. Three scales are privileged. While blatantly racial in South Africa, the administrative and legal apparatus of exclusion, implemented by the colonial and apartheid states, was, in urban areas, simultaneously based on residence, age and gender, and in both countries, re-appropriated and subverted by the population (Part I). Similarly, over the long twentieth century, order was maintained in low-income neighbourhoods more often by local organisations using coercion against those who appeared to threaten the cohesion of the 'community' than by the state security apparatus imposing us to rethink the very act of policing in low-income neighbourhoods (Part II). Eventually, what may be striking to the observer of the contemporary precarious urban life in Africa is their uneven ability to join clientelistic networks and negotiate the terms of their economic life with local political patrons and street-level bureaucrats (Part III). Each of the book's three parts focuses on a common scale of analysis and temporal period: the metropolis during the colonial period in Part I; the neighbourhood from the 1930s until today in Part II; and central sites of the urban economy (the street, motor parks) and of bureaucratic and political power (a local government office, the residence of a political boss) since 2000 in Nigeria.

In Part I, I suggest that the genealogy of exclusion takes root in the classifying obsession of colonial and apartheid governments. Racial discrimination, city planning and segregation schemes designed to keep black native populations at a distance from white European populations have played a pioneering role in developing the concept of a segregated city (Home 1997; Nightingale 2012; Swanson 1977). For American historian Carl Nightingale, it is impossible to wish away the power of race in the history of colonial cities, whether it was during moments when planners did draw colour lines clearly on their maps or moments where they sought to hide race away (Nightingale 2012). He rightly insists, 'the biggest problem with urban racial separation is the maldistribution of resources that disproportionally disfavours those racial groups that the colour line also helps to subordinate politically'. Actually, the construction of the modern networked city in the Western world from a fragmented provision of water, sanitation, electricity or transport in the mid-nineteenth century to a standardised system in the mid-twentieth century – or what is referred to as a 'modern infrastructural ideal' by Graham and Marvin (2001) – never happened in the colonies. Network of infrastructures has 'always been fragmented' in colonial Bombay (McFarlane 2008; Zerah 2008), it was a truncated modernity in colonial Lagos (Gandy 2006, p. 377) and even considered as a system of urban apartheid in the earlier literature (Abu-Lughod 1980; Balbo 1993). An approach on colonial surveillance and control of conducts has simultaneously become an important reinterpretation of the colonial city inspired by a Foucauldian analysis. Some have suggested that disciplinary colonial power and the racist narrative behind the civilising mission were penetrating the smallest details of everyday life that could only be resisted by colonised social and political practices (Celik 1997; de Boeck and Plissart 2004; Myers 2003).

While the racial classification and the surveillance dimension of the colonial project are too important features to be brushed aside, it does not fully account for the role of bureaucratic, political and social engineering in shaping colonial classification, creating exclusion and producing violence.

The analysis of colonial cities as dual cities or areas in which coercion was tempered solely by forms of resistance misses the essence of what thirty years of historiography has taught us about Africa. Colonial societies cannot be understood merely as antagonism between Europeans and natives (Cooper 1994; Eckert 2006, p. 213), which reduces the colonial encounter to binary opposition (elites/subalterns, domination/resistance) (Bayly et al. 2006, pp. 1452–1456).

The notion of the dual or racialised city overestimates the ability of administrations to implement steady public policy and ignore the dispersion of colonial power into a multitude of locales and authorities. It overlooks the capacity of local societies to circumvent, ignore or even conceive of such divisions differently and misses the pervasive social effects of other forms of colonialism in urban Africa.

First, the racial delusion of colonial power was articulated to other forms of classification based on residence, origin, gender and age, which together constitute a classifying obsession needed to govern urban areas. Colonial authorities were faced with contradictory injunctions. The first was the will to assign migrant populations to rural areas and the need for those same populations as manpower for the urban or the industrial economy. One of the solutions was to provide them with a specific place of residence and grant them different rights from the more permanent urban population even if this was difficult to implement as populations keep moving between urban and rural areas (what is referred to as the *population flottante* in French colonies). A second issue arose from the labour policies aimed at identifying and promoting an autonomous male working class, which became the standard against which a large part of the urban population became criminalised. 'Unemployed', 'idle', 'unruly' youth and 'single' women without wage jobs were increasingly seen as contributing nothing to the colonial economy, whereas their ordinary behaviour threatened the authority of chiefs, elders, husbands and wage workers. The invention of a nomenclature designed to rule by classification was central in governing the most dominated social groups in cities under the colonial rule.

Secondly, it is dubious to qualify the increasing forms of colonial surveillance over the urban population as a form of governmentality or in the words of Foucault as a process aimed to identify and reform individual conducts.[9] Historians of Africa are sceptical. Colonial domination was based less on the creation of individualised subjects than on the reified notion of traditional authorities (Cooper 2005). The colonial states in Africa did not necessarily need their subjects to be individualised and identified by the state and colonial institutions complained more about the lack of collective adjustment than individual adjustment (Cooper 2005). The process of counting the population for tax or demographic reasons, or weighing, classifying and measuring them in the case of workers did not attest itself to the emergence of a governmentality based on individualised knowledge of the population; it was a process of 'unitisation' rather than the creation of individual subjectivities (Vaughan 1991, p. 11). Far from being driven by a ubiquitous scientific curiosity about the well-being of the population, African colonial states were built in an informational void: there is an inadequacy of an account of the state motivated by the search for knowledge in colonial and post-colonial Africa (Breckenridge 2014, p. 5). Without immediately ruling out the analyses of governmentality, Frederick Cooper and Megan Vaughan sought to assess the relevance of this interpretive framework with regard to specific environments and time periods. The first part of the book tries to look at this specific surveillance and attempt to discipline new identified urban groups and determine whether specific *dispositifs* of government were able to identify and reform their individual conducts.

Thirdly, resistance to colonialism or apartheid is a very ambivalent process that could not be reduced to an opposition between an elite cooperating with colonisers against groups qualified as subalterns or in the words of Gramsci (1971) individuals subject to the activity of ruling groups. Moving away from a binary opposition between colonisers and colonised does not consist in underestimating the colonial violence but rather in thinking about the inextricable entanglement between the governing and the governed.[10] The new groups produced by the administration were not only abstract and fantasist bureaucratic categories but took roots in the urban social world. A detailed analysis of two notorious episodes of collective violence – the Kano riots in 1953 in Nigeria and the Sharpeville massacre in 1961 in South Africa – show not only resistance against colonialism or apartheid. Exploring the ways in which a set of various subaltern groups played different roles in these two episodes reveals how they could absorb many ideas from the ruling class while at the same time judging their everyday experience contradicted that domination. Administrative categories produced during the colonial period have actually been appropriated by a set of social and political actors and lasted after the end of colonial rule or the apartheid regime.

Part II retraces the genealogy of policing carried out by organisations in low-income neighbourhoods from the colonial period to the present. In the policing literature, there is a distinction between law enforcement carried out by the 'police' – the name commonly used to designate a state organisation with a specific mandate – and 'policing' which designates a plurality of organisations including the police (Garland 2001; Jobard and Maillard 2015). This second part focuses on groups and individuals policing neighbourhoods often included under the term 'vigilante'. It wishes to open up a nascent dialogue between comparative urban studies and the history and anthropology of policing to rethink the act of policing in low-income neighbourhoods.

The anthropology of vigilantism on Africa has, on the one hand, largely focused on the relationship between vigilante groups and the various arms of the state, their role in building communities, their use of violence and the multiple threats identified by vigilante groups (Buur and Jensen 2004; Kirsch and Gratz 2010; Pratten and Sen 2007; Smith 2019). Contemporary policing by non-state actors are to be found everywhere in Nigeria and South Africa but how this has emerged from specific struggles, alliances and conflicts remains unclear. There is actually a paucity of research on how the day-to-day work of these groups have reshaped ordinary urban experience over a long period of time, their role in exercising public authority in neighbourhoods, in excluding or including residents, and the ways in which they have participated in the regulation or banalisation of daily violence. A genealogical approach exploring what has been forgotten and what has been naturalised over time helps to reconsider the very act of policing in these cities. Joining historical and ethnographic methods enables answering key questions left aside in the literature on vigilantism. Is the violence manifested by these organisations the same when it is authorised (during the colonial period and apartheid) and when it is prohibited (today)? Are vigilante organisations regulated in the same way when bureaucratic resources are meagre (the colonial period) and when police forces have been significantly expanded (today)? How do they exercise power over the people residing there and who has the authority to govern the neighbourhood? Exploring past and present everyday policing helps to disentangle continuities and discontinuities from within the neighbourhood.

Vigilantism is sometimes understood as another form of neoliberal government (Comaroff and Comaroff 2006; Goldstein 2005). The delegation of security to private actors takes part in a move towards what David Garland rightly calls 'responsibilization' that is, the acceptance that individuals should be held responsible for their own security (Garland 1997, pp. 190–191). This well resonates with the neoliberal urbanisation argument in comparative urban studies. Since the 1990s, large metropolises have been identified as essential vehicles for the reproduction of neoliberalism – understood as a body of doctrines imposing the adoption of universal free market values. They are described as showcases for major macroeconomic transformations and privileged spaces for testing multiple schemes: areas granting tax exemptions, public–private partnerships, new strategies of social control and surveillance, multiplication of urban enclaves for the middle and upper classes (business clusters, shopping centres, gated residential communities and industrial parks) (Brenner and Theodore 2002; Caldeira 2000; Davis 1992; Peck and Tickell 2002). This privatization of urban space is closely connected to the geography of fortified enclaves while vigilantism is more especially associated with peripheral urban areas in which residents feel ignored by the legal state (Glebbeek and Koonings 2016, p. 8; Müller 2017; Rodgers 2006).

While I agree that 'urban neoliberal order' is marked by more systematic surveillance, the development of urban enclaves, and the delegation of security to citizens, it may miss other dimensions of urban policing in a poorer urban space that has not received the same scholarly attention (Jaffe 2012). Secured urban enclaves are common in wealthy Nigerian or South African neighbourhoods, but private security companies are rare or absent in low-income neighbourhoods and if residents are involved in community police programmes, in Nigeria and South Africa, such programmes have often taken over earlier systems of security mobilisation by local residents. The neoliberal urbanisation argument marginalises or ignores other forms of urban trans-formation (Le Galès 2016; Pinson and Morel 2016). In our cases, the longer colonial and postcolonial trajectories of urban vigilante groups indicate that vigilantism cannot be fully explained by neoliberalism or weak state analysis.

In articulating comparative history and comparative ethnography in two specific large urban areas in the cities of Cape Town and Ibadan, I inform how policing is the product of a very specific urban environment of police of subalterns by other subalterns. This specific genealogy has roots in the colo-nial and apartheid periods during which the administration delegated or 'dis-charged' (Hibou 1999) its security functions to very large number of groups and organisations at conditions that they did not challenge the overall colo-nial order. These groups were tolerated or supported by the administration but enjoyed a large autonomy. In many instances, policing the neighbour-hood often appears to be the other side of the classifying colonial obsession: youth, migrants or people unknown from the local residents were the main targets of vigilante groups. These power relationships between groups and those threatening the community have strongly persisted in the everyday rou-tine of urban policing. New unexplored issues have also come up since the end of the colonial or apartheid periods such as politicisation, commodifica-tion and feminisation of vigilantism. In other words, scrutinising daily anxi-ety in urban areas neglected by the state opens up new avenues for empirical and theoretical research on low-cost and harsh forms of urban policing.

Part Three moves from a genealogy of exclusion at the city level and police at the neighbourhood levels to *dispositifs* of power at the micro level on the streets and in office from the 1990s to date. It explores everyday relationships in bus terminals referred to as 'motor parks' and in local government offices between individuals in positions of authority (political leaders, civil ser-vants, trade union members) and a host of subordinate actors (bus drivers, tax collectors, unemployed workers, ordinary citizens seeking a document) in three main metropolises of Nigeria (Lagos, Ibadan and Jos). Focusing on these places offers an opportunity to analyse everyday practices of exclusion and inclusion in a clientelistic network, a political community, or access to employment and forms of violence that such an exclusion might trigger.

A world of a dominant urban precarity has become the norm in many African and Asian countries (Simone and Pieterse 2017, pp. 33–36). With the implementation of IMF policies in the 1980s, public sector retrenchment and cutbacks in social programmes and subsidies eroded the number of wage earners while making the lives of the majority of workers increasingly precarious (Barchiesi 2019, p. 68). In the face of these policies, state governments and even more so municipalities, that were ill-equipped to cope with the continent's demographic challenge, could not have prevented the growing exclusion of poor populations from basic services and the wage job market (Koonings and Kruijt 2009). Nigeria is no exception in that regard: the level of urban poverty and extreme urban poverty dramatically increased in the 1980s and 1990s.[11] Petty trading and petty craft have become the dominant sectors of employment for men and women in a city like Ibadan (Akerele 1997, p. 39; Guyer et al. 2002), transforming streets and public spaces into a major place to make a living.

In Nigerian metropolises like in many other cities of the south, relations between economic actors and state agents commonly labelled the 'informal economy' – usually refers to self-employment, i.e. either non-wage work or to activities that are not subject to taxation or state regulation – have acquired very different political meanings over time. Many studies on urban informality have emphasised the vitality of horizontal ties and tended to overlook vertical relations even if the literature has informed crackdowns, evictions and brutal harassment by local and national authorities (Lindell 2008; Potts 2008), stronger control of CBDs, sometimes by re-enacting old colonial measures (Morange 2015; Steck et al. 2013). Less has been said on hierarchical social relations within networks of informal actors and political leaders (Lindell and Utas 2012), the ambivalent politics of street trader organisations (Bénit-Gbaffou 2016), the fragmented mobilisations to access services or collective resources and urban spaces (Bayat 1997, 2000) and the complex interplay between politicians, urban population and their intermediaries in the city (Auyero 2000; Haenni 2005; Solomon 2008).

The politics of the street in Chapter 5 is not an additional chapter on urban informality but an attempt to account for the ambivalent role of bureaucratic and political domination in creating exclusion, in producing violence and uneven forms of politicisation. What is happening in streets and motor parks does not easily fit what a number of scholars would qualify as 'political' in its emancipatory dimensions. Following Weber (1978), Hibou (2011) and Lüdkte (2015a) it looks instead at politics as participating in power. This chapter explores how Lagos and Ibadan political leaders have used their networks to provide infrastructures and services to the urban population and how the thousands of motor parks in the two largest cities of the county have become central places of illegal tax collection. It scrutinises how subaltern actors in the parks participate to this collection and became politicised through conflicts and more daily practices. It tries more generally to understand the porous border between the legal and the illegal, state and non-state actors, the circulation of petty cash in and out of official circuits and forms of daily policing and violent mobilisation for the control of urban resources. Chapter 6 interrogates the politics in office in looking at daily interactions between civil servants and citizens seeking a certificate of *indigene* in local government front offices in Ibadan and Jos, Plateau state capital in the central part of the country. A quota system that was started in the 1980s guarantees the representation of citizens in the various administrations according to their origin certified by local governments. Certificates of *indigene* embody the official – in its original meaning, produced in office – and illegal discrimination in Nigeria. In Jos, this process of delineating who is truly indigene of the place have led to mass violence claiming several thousands of deaths in the 2000s. In this final chapter, I will examine conflict and negotiation required to procure this document, what such transactions tell us about the forms of inclusion and exclusion that are approved by the local administration, and the extent to which this practice is perceived as discriminatory by the *non-indigene*. I follow Matthew Hulls' suggestion that bureaucratic documents are both constitutive of bureaucratic activities and of forms of urban sociality that extend beyond the office (Hull 2012). The last chapter links this detailed topic of producing documents from within the office to wider politics of urban exclusion. Behind the politics of exclusion of *non-indigenes*, which is especially embodied in the graphic artefact of the certificate, there is a very unclear legacy of the past: in both cities, competing ancestral claims are rooted in the history of the city, in colonial urban planning and postcolonial administrations. Coming back to these different historical layers in a comparative approach helps to see the ways in which history has been used in more recent times as a political and bureaucratic tool to exclude *non-indigenes* to access limited state resources. Changing the scale of analysis in looking at detailed places of encounters such as motor parks and local government offices allows to observe ordinary social practices and mass violence, to interrogate the legacy of a colonial past and to explore the thin, blurring and often invisible borders between civil servants and unofficial tax collectors and the embeddedness of state bureaucratic practices in urban society.

Notes

1. A glossary at the end of this book explains specific terms with an asterisk following their first use in the text (author's note).
2. These two scales should be kept in mind because they suggest that processes of inclusion worked elsewhere in the two countries.

Fortunately, the vast majority of the townships in South Africa and cities in Nigeria did not experience this kind of tragedy.

3. Lagos was one of the main ports for slave trading on the Bay of Biafra in the nineteenth century, and the political capital of Nigeria from 1914 to 1991; Ibadan was the main city-state in western Nigeria in the nineteenth century, the capital of the western region from 1946 to 1976, and the capital of the state of Oyo since 1976; Kano has been the economic capital of the Sokoto Caliphate since the nineteenth century, the capital of the northern region since 1900, and the capital of the state of Kano since 1967; Johannesburg has been the world's capital of gold-mining since 1890, the economic capital of South Africa, the provincial capital of Transvaal from 1910 to 1994 and of Gauteng since 1994; Cape Town was the capital of the Cape Colony in the nineteenth century, the country's leading port, the seat of Parliament and the provincial capital of Western Cape since 1910.

4. An apparatus is a network comprising 'a thoroughly heterogeneous ensemble consisting of discourses, institutions, architectural forms, regulatory decisions, laws, administrative measures, scientific statements, philosophical, moral and philanthropic propositions – in short, the said as much as the unsaid' (Foucault 1977, p. 299 and for the translation Gordon 1980, pp. 194–228).

5. Based on the official censuses in these countries.

6. The research on ethnic identification in the twentieth century is dated, profuse and disputed (Spear 2003).

7. The volume of the sources is uneven, depending on the topics, which is why the sources used have been presented in detail in the introduction to each part.

8. *Evènement il faut entendre par là non pas une décision, un traité, un règne ou une bataille mais un rapport de forces qui s'inverse, un pouvoir confisqué, un vocabulaire repris et retourné contre ses utilisateurs, une domination qui s'affaiblit, se détend, s'empoisonne elle-même, une autre qui fait son entrée, masquée.*

9. 'By governmentality, I mean the ensemble formed by the institutions, procedures, analyses and reflections, the calculations and tactics that allow the exercise of this very specific albeit complex form of power, which has as its target population, as its principal form of knowledge political economy, and as its essential technical means apparatuses of security. Secondly, the tendency which, over a long period and throughout the West, has steadily led towards the pre-eminence over all other forms (sovereignty, discipline, etc.) of this type of power which may be termed government, resulting, on

the one hand, in the formation of a whole series of specific governmental apparatuses, and, on the other, in the development of a whole complex of savoirs' (Foucault 1978, p. 655, trans. Gordon 1991).

10. We are borrowing Gramsci's methodological suggestions here rather than closely following his thinking, which was inspired essentially by an Italian context not easily transposable elsewhere. For a use of Gramsci in African urban contexts, see Choplin and Ciovalella 2016; Glassman 1995.

11. According to the Federal Office of Statistics, poverty levels increased from 28.1 per cent in 1980 (representing 17.7 million people) to 65.6 per cent in 1996 (representing 67.1 million Nigerians) (FOS 1999, p. 24). If the extremely poor people group was not important in 1980 in urban areas (around 3 per cent), in 1996 it affected a quarter of the urban population (FOS 1999, p. 26).

PART I

Governing Colonial Urban Space

What do a temporary migrant, an urban dweller, a native, a non-native, a delinquent, a single woman, and a child in need of protection have in common in the context of two countries such as Nigeria and South Africa in the early twentieth century? In principle nothing, except for the fact that these terms came into being at a particular time (the colonial period) in a specific environment (cities) and were part of a nomenclature for the instruments of power designed to govern populations by classifying them into categories. The aim of Part I is twofold: first, to analyse the environment that generated and introduced policies to identify groups to exclude them from access to urban space and its resources (housing, work, leisure activities) and second, to reveal the new social differentiations these policies produced.

The history of colonial domination has often been read as a profoundly violent and racist endeavour that robbed colonised people of their own history. But the internal contradictions of colonialism, its bureaucratic inefficiency, negotiation, mutual instrumentalisation and the capabilities of the colonised to take action have made it impossible to reduce this domination solely to its violence or its surveillance functions, including in South Africa.[1]

In this case as in others, the exercise of domination implied taking advantage of its 'insidious gentleness' or *douceurs insidieuses*, to employ Michel Foucault's expression, recently used by Béatrice Hibou (Hibou 2011), i.e. the observation that it was often in the concrete, circumstantial and historically situated moments that the day-to-day practices of power and the ambivalent relationship between rulers and the ruled were manifest. Adopting such a perspective in the colonial context frees us from thinking in terms of oppositions such as collaborators vs opponents, racial coercion vs African resistance and the elite vs the people. For example, the forms of cooperation between townships, elders and apartheid police in security matters and the pragmatic acquiescence of Africans to the social housing programmes of the apartheid government (Evans 1997, pp. 155–159) cannot be understood as simply offsetting authoritarianism by social programmes (Hibou 2011). Similarly, contemporary – or colonial – bureaucratisation can be seen not so much as the result of strategically motivated public policies, coherent strategies or control and subjugation alone, but as the product of a constellation of interests – to use a Weberian expression – that unfolds through the many different actors who are stakeholders in this process (Hibou 2013, pp. 13–14).

An urban environment offers an excellent vantage point for observing the way such bureaucratic procedures and routines develop. Empirical knowledge about the so-called urban populations of Africa expanded under the impact of social and political changes in the first half of the twentieth century. Until then, the colonial and South African authorities had been steeped in a dual vision of the continent, opposing the primitive African to the civilised European, the pre-capitalist to the capitalist, the tribal system perceived as rural and the urban world perceived as modern – in which urbanisation was seen as a process of modernisation (Cooper 1983; Ferguson 1999, p. 86). Indirect rule[2] in the British Empire, racial policy in French West Africa[3] and the reserve system in South Africa were all inspired by the idea that native populations should be kept in place to make it easier to control and govern them (Geschiere and Jackson 2006, p. 4). In the eyes of the administrations, cities were created to meet the needs of the imperial economy or the mining industry were foreign to the practices of African people; they imagined rural societies in which people respected the authority of the chiefs and elders, and opposed them to a menacing urban world where detribalisation was under way (Burton 2005; Lewis 2000). The south-western (Ibadan, Lagos, Abeo-kuta) and northern city-states (Kano, Zaria, Zamfara) that predated the conquest of Nigeria, were also viewed and administered as traditional spaces, according to the principles of indirect rule.

Very soon, however, the colonial and South African authorities were faced with contradictory injunctions. How can migrants be assigned the role of making the colonial or industrial economy work without calling into question racial hierarchies in South Africa or the native authorities in Nigeria? And what should be done about 'unemployed' youths, who were 'idle' or 'unruly' and 'single' women without wage jobs, whose ordinary behaviour threatened the authority of the chiefs, elders or husbands? One of the solutions the colonial and South African authorities found to these problems was to identify individuals and groups and manage their differences according to geographical or ethnic origin, sex or age.

The analysis we are proposing here adds complexity to the definition suggested by Mahmood Mamdani (Mamdani 1996) of a bifurcated colonial state resulting from the institutionalisation of two systems of power by the colonial authorities: the first, an urban, racialised system, based on rights guaranteed to Europeans; the other, rural and tribalised, based on the despotism of customary law and chiefs described as decentralised despots. Mamdani's analysis, according to which African workers and middle-class urban residents found themselves in a legal vacuum (Mamdani 1996), schematically opposes urban and rural areas, whereas urban Africans also came under the authority of traditional chiefs. It fails to mention that colonial authorities were continually implementing specific legal apparatuses for Africans residing or working in the city; and it is a de-historicised analysis as the categories of urban dwellers and workers were continually shifting in labour policies (Cooper 1995). More broadly, this book asks whether urban spaces were areas for testing and implementing a governmentality aimed at reforming individual conduct. Collected data by colonial administrators did not automatically lead to the implementation of specific policies – far from it – and one must be circumspect about their performative aspect (Bonnecase 2011, p. 21). Rather than the product of a coherent machinery, measures implemented by colonial power were makeshift devices that grew out of fragmentary empirical knowledge, the concerns of the moment, the expansion of the agencies or ministries in charge of these populations and the support of certain groups who saw them as furthering their own interests, all of which could, in the process, alter the initial objectives.

Colonial cities offered spaces to develop and implement labour policies and population management, relying on the exclusion of differentiated groups according to the context and the period – exclusion that may have pertained to temporary migrants, but which also encompassed other more or less sizeable categories or subaltern groups (the unemployed, single women, non-natives, delinquents, street traders, etc.). Our purpose here is not to write the social history of these groups, nor to give voice to those once reduced to silence by their conquerors. Instead, we aim to show that the metropolis was a space that simultaneously produced social subordination and constructed a bureaucratic reality. This task implies thinking about the inextricable entanglement between the governing and the governed, and questioning the extent to which the various instruments of power that were introduced generated new social realities and participated in building a bureaucratic state that devoted much more time, energy and resources to gathering information on these people, classifying and categorising them and allocating reserved spaces to them.

We will focus on certain measures that favoured the exclusion – and inclusion in urban space – of the following specific groups: temporary migrants (versus urbans) in South Africa, non-natives (versus natives) in Nigeria and young delinquents (versus adults) in both countries. At the very core of the processes of social differentiation, these measures were not only imposed from the top by the bureaucratic apparatus, they were also sponsored, shared and assimilated by a number of actors favourable to these policies. Thus, specific urban policies were developed for particular age groups identified, depending on the context and the period, either as dangerous (juvenile delinquents) or endangered (children in need of care). Though they were carried out by social services seeking to reform the conduct of children and youth, they were also supported by elders and parents in townships who saw these young people as threat to a certain moral view of the world (Chapter 2).

These policies were simultaneously appropriated and challenged by the population, which indicates their relative importance. The Kano riots in 1953 and the Sharpeville massacre in 1960 revealed new social configurations that originated in the colonial domination (Chapter 1). They show that the marginalised groups express their own demands, thereby demonstrating autonomous political initiative, but at times within the horizon of the language and practices of government. They also bring out divergences between the social and political interests of the various groups involved. Our analysis thus marks a break from an approach centred on nationalism and resistance to colonialism or apartheid. Similarly, it departs from a class-based approach that sees the marginalised population either as dominated by elites and incapable of revolt or as rebellious and driven by awareness of their precarious economic situation.

Notes

1. Regarding these two successive approaches, see Achille Mbembe (2010) and Vincent Foucher's (2010) answer. See also the analysis in terms of hegemonic transactions developed by Jean-François Bayart (1989) and on South Africa by Deborah Posel (1991).

2. Indirect rule was a way of exercising colonial power by governing through native leaders, who were given power to deal with legal matters, policing, administration and taxation and overseen by the British authorities. First implemented in India, and experimented in Uganda and later Nigeria in the early twentieth century, indirect rule became the official policy of the British Empire and was extended to its other African colonies during the interwar period.

3. A race-based policy, promoted by the governor general of French West Africa (AOF), William Ponty, starting in 1909, which relied on chiefdoms to act as intermediaries between the colonial administration and ethnic groups or races henceforth codified in an official taxonomy.

CHAPTER 1

Classifying and Excluding Migrants

Race, residence and origin were the three criteria defined by the colonial administration of the Nigerian and South African governments that were used to create systems of exclusion and inclusion in the urban areas of the two countries. The first criterion related to the administrative racial classifications then in force: blacks, whites, coloureds and Indians in South Africa; Europeans and natives in Nigeria.[1] The second criterion separated certain groups in South Africa described as 'urbans' or permanent residents, who were granted the right to live in the city, from the vast majority of rural Africans classified as 'migrants' or 'temporary sojourners', who were deprived of such residential rights. The third criterion dissociated groups in Nigeria, who were said to be 'of native origin', from groups known as 'non-natives' or 'foreign natives'. A considerable amount of research has been devoted to the definition and evolution of these classifications as well as the social reality they helped to produce, but these three aspects are usually treated separately in the literature.

The first criterion was linked to the question of racial classification, theoretically modelled on the differences between citizens and subjects, but in fact difficult to dissociate from socioeconomic and cultural criteria, a reality attested by the obsession with how to classify 'borderline' racial groups such as mixed-race both in colonial Africa (mixed blood) and in South Africa (coloured) (Martin 1998; Saada 2007). On the scale of the city, this criterion was also associated with the racialisation of urban areas, i.e. segregated residential policies based on belonging to one of the official racial classifications, an issue that has long been at the heart of research in urban history. Studies of colonial city planning and segregation schemes designed to keep native populations at a distance from white European populations have played a pioneering role in developing the concept of a segregated city (Home 1997; Nightingale 2012). The concept evolved in two different and unexpected directions during the

Classify, Exclude, Police: Urban Lives in South Africa and Nigeria, English Language First Edition. Laurent Fourchard.
© 2021 John Wiley & Sons Ltd. Published 2021 by John Wiley & Sons Ltd.

1990s: the first, centred on South Africa, stresses the need to transcend the racialisation of urban areas, and the second, centred on cities in the rest of Africa, insists on the concept's continued relevance, despite criticism about this dualistic interpretation (Bigon 2016; Fourchard 2011a). We would like to go back and examine the reasons for these divergent historiographical shifts, which were surprising to say the least, given that the exacerbated racial division of urban space imposed in the 1950s was still a very tangible reality in South Africa in the 1990s.

The second criterion – residence – referred to the roles assigned to Africans in urban areas by labour policies, above all the distinction between the so-called temporary or 'floating' populations (circulating between the city and the countryside) and the so-called 'stabilised' populations (i.e. a working class with its own system of wages, promotions and status, which developed family life by bringing workers' wives to the city and putting them in charge of the children's education) (Cooper 1995). In Nigeria, stabilisation policies were introduced at a late stage after the Second World War, as in the rest of British and French colonial Africa. These policies were limited by the centrality and size of the non-wage sector as well as urban traditions that predated colonisation and the fact that wage earners set up family networks extending far beyond the nuclear families fantasised by the colonial administration (Lindsay 2003). In South Africa, the labour market and the divisions between migrants and families in townships were structured over the long term by the coexistence of policies granting different rights to workers depending on whether they resided in urban or rural areas. Jeremy Seekings and Nicoli Natrass have emphasised the extent to which the current social inequalities in South Africa are the legacy of this long-standing differentiated management of the population (Seekings and Natrass 2006). Such policies radically transformed urban environments but also directly shaped the South African bureaucratic apparatus, and aggravated the precariousness of the country's most vulnerable groups, the migrants from the Bantustans whose rights to residence in the city and thus access to employment were always under threat – a process that was challenged during the demonstrations and riots that took place in Sharpeville and Langa in March 1960.

The third aspect of classification – origin – is related to autochthony, i.e. the claim to have been the first to arrive in a locality (Bayart et al. 2001). Autochthony is inherent in the formation of the colonial and post-colonial state, which in seeking to govern through determining the location of African residents redefined the classifications of 'subject' and 'citizen' as well as the boundaries between those who 'were there before' and those who came afterwards (Bayart et al. 2001; Geschiere 2009). Research on autochthony in Africa has concentrated above all on rural property conflicts (Dorman et al. 2007; Kuba and Lentz 2006; Lund 1998), war situations and projects to achieve political hegemony (Banégas 2006; Gary-Tounkara 2008; Marshall-Fratanti 2006). Far less attention has been given to the urban dimensions of these phenomena, with a few notable exceptions, including in Nigeria, a particularity that deserves a closer look (Albert 1996; Akinyele 2009; Anthony 2002; Cohen 1969; Fourchard 2009; Higazi 2007; Madueke 2019; Mustapha and Ehrhardt 2018).[2] 'Indigeneity' is the standard term employed by actors and researchers to describe autochthony in this country. During the 1970s, the terms 'indigenes' and 'non-indigenes' probably replaced the words 'natives' and 'non-natives' used during the colonial period (see Chapter 6). At the time, colonial management of population flows was less concerned about dissociating temporary migrants from permanent urbans than about preserving the authority of 'traditional chiefs' by removing their responsibility over migrants coming from other regions of the federation. The colonial administration had decided to grant some of these migrants a particular place of residence (reserved neighbourhoods) as well as institutions and rights that differed from those of the local population, thereby creating a lasting distinction between populations of native origin and those said to be non-native. This single device was at the core of the process of differentiating the two categories and the particular historical forms indigeneity took in Nigeria. In the highly competitive political context of decolonisation, the rivalries between these groups turned into conflicts, like the one observed in the Kano riots in May 1953.

The politics of exclusion and belonging based on race, residence and origin are often treated separately on a national, regional or imperial scale. They are harder to dissociate on the scale of the city seen as a totality to be governed by an administration despite the presence of multiple competing authorities in such areas. The excessive focus on the racial dimension sometimes drove other social realities out of the spotlight, though they were in fact part and parcel of the same colonial project. The laws passed in the early twentieth century to organise racial segregation in urban areas (the decree of 1917 in Nigeria and the law of 1923 in South Africa), also governed residence and guaranteed rights of 'natives' and 'non-natives' in Nigeria as well as those of 'urbans' and 'migrants' in South Africa. The racial question was inextricably linked to urban and rural residential policies, which played a central role in South Africa's mining and industrial economy,[3] while the question of native or non-native origin was essential to the smooth functioning of indirect rule in Nigeria. It was also inseparable from the emergence of categories endowed with specific rights ('urbans' in South Africa and 'natives' in Nigeria). Those categories laid the legal foundation for the exclusion from the city of all those supposedly outside their definition ('migrants' in South Africa, 'non-natives' in Nigeria). On the urban scale, this resulted in assigning distinct areas to groups (usually specific locations, townships or hostels), granting them specific rights in terms of housing and work, placing them under a distinct authority, in some cases guaranteeing privileges or exemptions and facilitating access to services such as education, health care and leisure activities.

To understand more fully how these different policies of exclusion and belonging interacted in the two countries, I will examine how the policies came into being, tracing the initial voluntary participation of those concerned to their violent opposition to such policies, in particular by looking in some details at two local events that drew international attention: the Sharpeville massacres in 1960, viewed by some as proof that half a century of non-violent opposition had failed and that South African liberalism had been defeated by radical movements (Rich 1986); and the Kano riots in 1953, often described as the first communal riot in Nigeria. More specifically, in line with Michel Dobry, I want to look closely at what actually took place in these moments of crisis (Dobry 1986). The aim is to articulate different historical durations in the analysis: how the manufacture of differences between groups were co-produced by the bureaucracy and local populations in the medium term; the shorter-term history of the three places (Kano, Sharpeville and Langa), which must be re-examined in detail precisely because of the local nature of the massacres; and finally, the very short duration of the event itself, how it unfolded and the actors it mobilised (versus those who stayed on the sidelines).[4] Taking these different historical periods and scales into account will bring out more clearly the meaning of the events especially in relation to the contested politics of categorisation and exclusion of migrants from the city.

Race and Urban Space

Although the story of racial segregation in South Africa and Nigeria is well known, it is important to remember that the measures implemented in these countries were incomplete. They were also on opposite trajectories by the 1940s, when race was becoming a marginal issue in Nigeria, but turning into a constant obsession of the South African government. This approach is especially necessary in light of recent historiographical interpretations regarding the role of race in shaping colonial cities.

At the end of the nineteenth century, urban development in both contexts was shaped by a powerful medical or sanitary discourse. A stronger correlation had been demonstrated between mosquito-borne tropical fevers and native districts, and it became urgent to remove their populations and keep them away from European neighbourhoods. Several historians have shown that fevers and epidemics often triggered interventions by government or municipal authorities, who deliberately used the argument of sanitary conditions to justify wider social and spatial separation between Europeans and

Africans (Bigon 2016; Goerg 1997; Ngalamulume 2004; Swanson 1977). The biomedical reconfiguration of urban space also reinforced differences among African residents themselves, even though, outside of epidemic episodes, such reconfiguration was more often a goal than a reality (Echenberg 2002). Measures excluding the entire African population from reserved areas were always partial, while many Africans managed to circumvent colonial rules (Bissell 2011).

In Nigeria, there was no legal foundation for residential segregation prior to the creation of European reservations in 1915 and the adoption of the law on townships in 1917 prohibiting Europeans from living outside their assigned districts and non-Europeans, except for domestic servants, from living in the European district (Olukoju 2003). The law replaced the term 'cantonment', designating areas allocated to quartering troops, by the term 'township', which in this country came to mean an enclave outside the jurisdiction of native authorities, intended to accommodate Europeans or so-called 'non-native Africans'. In practice, however, the clear-cut delimitation of a district exclusively for European use proved difficult to achieve and was not even desired by the various sections of the European community. In the early 1920s, the 1917 law was modified to make it applicable to the particular features of each city: strict residential segregation was no doubt possible in the new townships that had no long-standing European commercial interests (like those in Kano or Enugu). The law was considerably relaxed in older cities such as Lagos, where it simply recommended that a European zone be established where officials should reside, and other Europeans were welcome – but not required – to live (Olukoju 2004). The aim was to reconcile the interests of medical health officers with those of European traders and missionaries who refused to transfer their properties to European districts. This explains why the historic section of Lagos Island, in which European and African trades-men had been living and working since the nineteenth century, remained a mixed district. In 1928, Ikoyi, a residential neighbourhood in Lagos, was nev-ertheless planned exclusively for British civil servants. In the end, for a variety of reasons – medical progress in fighting yellow fever and malaria, the desire of the Colonial Office to abandon the principle of racial segregation in the mid-1930s, and the struggle engaged by nationalist leaders against any form of racial discrimination – most European districts in Nigeria like Ikoyi were turned into government residential areas (GRA) in 1940, where social status took precedence over racial origin after World War II (Olukoju 2004). In terms of city planning, the initiation of urban development projects involving racial segregation lasted only three decades in Nigeria.

In South Africa, the racialisation of urban space dates further back – the first location reserved for Africans is said to have been mapped out in Grahamstown during the 1840s and in Port Elizabeth in the mid-1850s

(Baines 1989). The policy was adopted more widely in the late nineteenth and early twentieth centuries where bubonic plagues have exercised a powerful influence on the development of urban segregation: areas were reserved for Indians in Durban and Johannesburg and for Africans in Cape Town and Port Elizabeth in 1901 and in Johannesburg the following years (Baines 1989; Parnell 2003; Swanson 1977) and was officially kept in place until the 1980s. The meanings given to the words 'location' and 'township' differed noticeably from those in Nigeria: in South Africa, they designated an area reserved for non-white people, separated from the rest of the city and usually situated on the outskirts. The first legislation regulating the residence of Africans in urban areas (Native [Urban Areas] Act of 1923) aimed to harmonise disparate local laws. It empowered municipalities to set up segregated locations outside the white city and elect advisory boards to introduce a budget and regulate the flow of migrants. The act was intended as a first step towards solving the problems of overpopulation, epidemics and criminality found in poor areas and sought to improve the living conditions of the African population through housing programmes provided by municipal governments, employers or Africans themselves. During the interwar period, the cities that were not reluctant or opposed to the idea were nevertheless unable to manage population flows and segregate population groups completely.

When the National Party came to power in 1948, it had already adopted 'apartheid' as its slogan (meaning 'separate development of the races' in Afrikaans), but it had yet to come up with a plan for its implementation (Posel 1991). Preserving white supremacy entailed a commitment to racial purity, the refusal to grant political and social rights to the African majority, and quartering people according to their racial classification. Racial planning became more systematic, relying on a new legal arsenal that imposed race-based residence, denied the possibility of ownership to the African population (Group Areas Act of 1950) and racially segregated access to public facilities, services and offices, from maternity hospitals to cemeteries, as well as cinemas, beaches, public parks, transport, places of worship and playing fields (Reservation of Separate Amenities Act of 1953). Although there were some areas still free from state control in the 1950s, the strict segmentation of urban space was a fundamental feature of apartheid. Today, the correlation between space and race remains the most obvious legacy of urban administrators in the apartheid era (Evans 1997, p. 121) as shown by ongoing efforts at racial integration, which was still restricted to the African 'middle classes' in the 2000s and the powerful heritage of racially divided urban space (Lemon 1991; Seekings 2000, p. 834; Tolimson et al. 2003).

It is perhaps because racial segregation was taken to such an extreme in South Africa that an approach emphasising the excessive racialisation of urban history came under fierce attack as soon as apartheid ended. In the mid-1990s, Jennifer Robinson and Paul Maylam criticised South African historiography for being obsessed with the racial question and suggests instead to analyse townships as units of state power (Robinson 1996) or to explore the relationship between the central state and local authorities in developing urban policies (Maylam 1995). Deborah Posel (1997) and Maylam rejected the teleological, monolithic and functionalist analysis of state control for creating the misleading impression of all-powerful state bodies. In the same issue, Susan Parnell and Alan Mabin criticised historians for exaggeratedly racialising their research topic instead of looking at how municipalities and city planning had been gradually introduced under the influence of the transnational movements in architecture and urban development, thereby ruling out comparative approaches between South Africa and the rest of the world (Parnell and Mabin 1995, pp. 39–61). In other words, for a number of South African scholars, apartheid was not only about state racism or the product of a single 'grand plan' created by the state in response to the pressure of capital accumulation (Posel 1991) but also about the significance of African townships as a key institution of segregation and resistance (Bickford-Smith 2008; Bonner and Nieftagodien 2008; Maylam 1995), as a strategy between spatial arrangements and political power (Robinson 1991) and at the centre of conflicts and compromises between the government and capitalist interests (Posel 1991). More recently, Achille Mbembe and Sarah Nuttall came to a similar conclusion: most historical studies of Johannesburg have been concerned with spatial dislocation, racial polarisation, evictions and the marginality of townships, and consequently neglected the social interconnections between the city, the townships and other areas (squatters' camps, the countryside, homelands) and underestimated the practices and imaginaries of townships and the role of townships in producing multiple urban identities (Nuttall and Mbembe 2009, pp. 356–357). Instead of being open-minded towards worldwide literature on the metropolitan experience, the history of Johannesburg was caught between two different teleological narratives: the rise of apartheid and the rise of the nation-state (Mbembe and Nuttall 2005, p. 198).

Just when a racialised reading of the apartheid city seems less central to South African urban studies, it is becoming prevalent in certain analyses of colonial and post-colonial cities. The notion of 'post-colonial' cities implies it is necessary to understand the colonial legacy they may share to include Africa's cities in a global framework suggested by the promoters of urban comparison (Robinson 2011; Simone 2010). The expression nevertheless conveys something more essential than a given historical moment, whether it expresses a reality radically different from the colonial city or on the contrary, a legacy that the post-colonial city cannot seem to eradicate. But in these new analyses, the 'colonial city' is not considered a total social object, as Georges Balandier (1985 [1955]) proposed; instead the focus of attention has been on urban planning, control technologies, the civilising mission and the separation between colonisers and the colonised and discussion of African social practices refer to a colonial perception of disorder and a resistance to colonial order (Celik 1997; Home 1997; Nightingale 2012). In his book on Kinshasa, Filip de Boeck suggests that colonialism produced a dual city (a native city versus a Western city) that dreamed of fashioning nuclear families of working class Africans, well dressed, well groomed, well fed and sexually domesticated (De Boeck and Plissart 2004, pp. 20–40), though he fails to mention the inadequacy of colonial policies to make this possible. He notes that it took 'the development of new peripheral areas, long after independence, to move away from the mimetic reproduction of the alienating model of colonial modernity imposed by the colonial state and Mobutu' (De Boeck and Plissart 2004, pp. 20–40). Resistance to the colonial order was perceptible in the youth movements of the 1950s that combined music, theatre and cinema to challenge these forms of imprisonment. In short, Belgian colonialism is said to have produced a segregation of lifestyles while violently penetrating the private, everyday lives of workers, a reality that the youth of Kinshasa more or less successfully resisted. On this account, the all-encompassing nature of the colonial project was limited only by the ability of the colonised to resist that same project.

Colonial societies cannot be understood merely as antagonism between Europeans and natives (Cooper 1994; Eckert 2006, p. 213). The notion of the dual or racialised city overlooks the ability of local societies to circumvent, ignore or even conceive of such divisions differently and overestimates the ability of administrations to implement a consistent, steady policy (Fourchard 2011a). It underestimates the internal cleavages especially between first comers and newcomers (Geschiere 2009; Locatelli and Nugent 2009; Monson 2015; Nieftagodien 2011) or the capacity of insiders and outsiders to develop a common sense of belonging to a town (Fabian 2019; Glassman 1995; Ranger 2010) the will of neglected communities to build an everyday alternative political order (Stacey 2019) or the wish of clandestine builders to blur distinction between the European city and the African 'suburbs' that had long shaped Lourenço Marques, today's Maputo (Morton 2019, p. 116). It also miscalculates the capacity of workers to escape forced or waged labour (Eckert 2019; Fall and Roberts 2019) and to enjoy new forms of urban leisure despite colonial restrictions (Fair 2018; Gondola 1997; Martin 2002). In the end, the point is not so much to de-racialise at any cost the urban history of Africa – and of South Africa in particular – but rather to step back and explore the new social configurations that have emerged from the colonial context and from apartheid. This, as I now consider, while the Sharpeville massacre is conventionally viewed in South African historiography as an especially significant manifestation of 'African resistance' to apartheid, it can also be reinterpreted as the product of policies to differentiate and exclude migrants from urbans and as a partial mobilisation against those policies.

Differentiating Urbans from Migrants in South Africa

On 21 March 1960, in front of the police station in the Sharpeville township (in the province of Transvaal, 70 km south of Johannesburg), the police fired on a crowd of demonstrators, mostly women, campaigning in favour of the abolition of pass* laws, i.e. the rules concerning an identity document required to legally reside and work in the city: 69 people were killed and 300 wounded. The demonstration was organised by the Pan Africanist Congress (PAC), a political party set up in 1959 following a split within the African National Congress (ANC). On the same day, in Cape Town, a crowd of between 6,000 and 15,000 people, all men this time, who had rallied in response to a call from the PAC in the township of Langa. The police gave the crowd three minutes to disperse before attacking, killing three individuals and wounding at least 46 others. That evening, the PAC activists set fire to the township's main administrative buildings, notably the entry and exit checkpoints for migrant workers. The army had to intervene to restore a fragile calm in the days that followed.

What made the Sharpeville massacre such a pivotal event in South African history was not the intensity of the violence – it was not the first time the South African police had opened fire on a peaceful crowd – but rather the political reaction it aroused within the country and around the world (Lodge 2011, p. 26). In South African historiography the Sharpeville massacre is analysed in nationalist terms and as the struggle against apartheid. According to Philip Frankel, Sharpeville marked an emotional high point in the anti-apartheid struggle and a decisive step on the long road from authoritarianism to democracy (Frankel 2001). Most approaches classify the events of Sharpeville under the heading of resistance to apartheid, which runs the risk of underestimating the weight of political divisions among the various anti-apartheid movements or ignoring conflicts between social groups that both contributed to the fragmentary character of the struggle (Frankel 2001; Legassick 2002). Three complementary approaches can be used to reveal this fragmentary character.

1. Gramsci argues that the state imposes its rule less by shaping people's consciousness than by the way it invites them to raise questions and debate certain topics. The events in Sharpeville and Langa unfolded in the context of specific measures taken by the South African government, namely, the policies that differentiated temporary migrants from urban dwellers. In the townships of Sharpeville and Langa, the radical disruption of lifestyles in the 1950s associated with a hardening of segregation into apartheid was predominantly conditioned by narrowing access to the city for male and female migrants. Once a central issue in the historiography of South Africa, this has been overlooked in recent analyses regarding Sharpeville. It is therefore necessary to re-examine this history.

2. In keeping with the ideas of Gramsci, who held that popular consciousness is not simply dominated by the thinking of the ruling class, but rather made up of fragments of thought from different periods and diverse horizons: the dominated can absorb many ideas from the ruling class, while at the same time judging their everyday experience as contradicting that domination (Choplin and Ciovalella 2016; Glassman 1995; Gramsci 1991). Taken together, those experiences enable us to understand more fully the actors' demands and grievances and highlight the constellation of contradictory interests involved, from the standpoint of both the decision-makers (ministries, companies, municipalities) as well as the mobilised actors (migrants, women, young students and delinquents, families). It is important to highlight these contradictions.

3. Adopting a detailed approach is the only way to deal adequately with this kind of history. Indeed, the demonstrations belong to a history peculiar to these two townships. Apart from Langa and Sharpeville, there was very little support in the country for the PAC's national mobilisation campaign against the pass that occasioned the violent incidents we will explore here. Thus, local contexts substantially determined the mobilisations. This means it is necessary to re-examine this highly local history in detail as well as in a wider context.

Stabilisation Policies and Urban Residential Rights

At various historical junctures, South African governments faced a choice between refusing Africans any legal status in the city and hence reinforcing their exclusion or building infrastructures likely to socialise them and acculturate them to the European vision of urban society (Cooper 1995, p. 515). In reality, these tendencies coexisted, and the focus was instead on the respective percentages of permanent residents and migrants that could be accommodated in urban areas. Employers wanted not so much to choose between a migrant workforce and a permanent workforce as to have access to both types of workers and balance their needs by playing one group off against the other or against the white working class (Cooper 1995, p. 264). Dual labour policies dissociating the two groups were adopted to meet these different requirements: one category of workers described as 'urbans' were to be stabilised by conceding them a few rights, whereas the population of temporary migrants were destined to return to the reserves.* This approach was less the product of a coherent policy than the fruit of compromise, negotiations and permanent conflicts within the government and power struggles between the bureaucracy, companies and white farmers in rural areas faced with a constant need for manpower until the late 1960s (Posel 1991).

Urban segregation in South Africa was grounded in the law of 1923. This law has often been presented as the legal basis for a system imposing temporary city residence on African population (seen as temporary sojourners) and encouraging the formation of a rural workforce sent back to the reserves once their labour contract was completed, in keeping with a model inspired by workforce management at the Kimberley and Johannesburg mines since the end of the nineteenth century. However, the law sought to widen the separation between temporary migrants and urbans, as Doug Hindson has shown (Hindson 1987). On the one hand, it recommended establishing a national system for inspecting work documents (passes) to control the flow of migrants more effectively. Municipalities were given the power to designate 'proclaimed areas' – most of the time there were locations and townships – where they could control population flows and repatriate anyone considered undesirable. On the other hand, it wanted to promote the stabilisation of certain categories of workers by granting pass exemptions to voters, property owners and professionals including businessmen, craftsmen, Africans with some education and employees. Women whose husbands were urban residents did not need a pass, a concession made to political organisations that had been hostile to such measures since the end of the nineteenth century (Hindson 1987, p. 36). In so doing, the law created a distinction between the urbans a category now defined legally who benefitted administratively from exemptions, and the other categories of the African population. While protecting the first group, they consolidated at the same time the power of local authorities to expel new categories of undesirables such as 'unemployed' people, i.e. those who failed to find work within two weeks after their arrival in the city, as well as single women and 'those who were idle, depraved or troublemakers' (Hindson 1987, pp. 39–41). The law conferred the right to punish job seekers in the so-called proclaimed areas, reflecting a new national preoccupation with sending the 'surplus' of African people back to the reserves, and systematising the use of passes to exclude the migrant population from access to the urban labour market (Hindson 1987, pp. 41–42).

While the law of 1923 intended to widen the separation between Europeans and Africans and dissociate migrants more completely from urbans, its application was hampered by a host of diverse concerns that made it impossible for municipalities to block the influx of rural populations seeking work in the areas under their authority.[5] Urban management of labour migration was problematic due to the many ambiguous and conflicting interests at stake between white employers and the central state, between the central administration and local authorities, and even within the Ministry of Native Affairs where the staff itself was divided on which policy should be adopted (Evans 1997). The mobility of the population was in the hands of hundreds of municipalities that jealously preserved their autonomy. In many cases, urban authorities decided not to establish proclaimed areas because it would mean paying for the cost of additional personnel to control population flows and they left the population living at the margins of the townships. They also lacked the resources to carry out ongoing territorial surveillance to deliver work permits to migrants, making it difficult to expel 'unemployed' workers when their 14 days of authorised residence expired. They refused to follow the injunctions of the Ministry of Native Affairs, while requesting additional financing for housing programmes. Lastly, until 1937 (when they were required by law to publish biannual municipal censuses), they had no reliable estimates of the urban population due to the existence of numerous squatter* areas and an unknown number of tenants and subtenants living in the townships (Evans 1997, pp. 45–49). In short, the knowledge acquired by the South African state concerning African urban populations was extremely fragmentary, whereas the dilution of responsibility between the central administration and the municipalities encouraged squatting in outlying areas and overcrowding in townships, which ultimately led to a major housing crisis at the end of the Second World War.[6]

By 1948, the preservation of white supremacy found itself threatened by an urban population responsive to the slogans of the ANC. The new forms of mobilisation adopted by the ANC since 1947 (demonstrations, boycotts and strikes) were a source of concern for the newly elected National Party. 'Tightening state power over the townships became a priority for the National Party and provided the incentive for building the apartheid system' writes Deborah Posel (Posel 1991, p. 270). Although, as Posel points out, the relatively weak majority of the National Party during this decade forced the government to make compromises, at the same time, the 1950s marked a significant break from the previous management of population flows in the city. In the late 1950s, Hendrik Vervoerd, Minister of Native Affairs from 1950 to 1958, then Prime Minister from 1958 to 1966, endeavoured to turn a divided Native Affairs Department (NAD), marginal in the state apparatus and focused on rural populations, into a key player within the South African bureaucracy, with added human and financial resources and a mission to make urban affairs a priority. He introduced new methods for controlling urban and migrant populations, reduced the autonomy of local authorities despite opposition, and significantly increased state information on the population, particularly through the reference book,* which gradually replaced the pass starting in the mid-1950s (Evans 1997, pp. 90–108). In 1959, the reference book became instrumental in implementing the prime minister's vision of 'grand apartheid', i.e. the systematic deportation of 'surplus people' to ten Bantustans. The Bantustans were 'reserves' set up on 13% of national territory that served, under the guise of self-government, as the main areas of South African rural ghettoisation (Breckenridge 2005, pp. 102–104; Giliomee 2003, p. 534). In apartheid South Africa, controlled urbanisation went hand in hand with bureaucratic expansion.

The aim was to freeze urbanisation by gaining greater control over migrants' access to the urban labour market. To meet the rising demand of white employers for manpower to fuel post-war economic growth, Verwoerd wanted to put the urban population to work by enacting the *Urban Labour Preference Policy* (ULPP), which gave urbans priority over migrants for jobs. The policy was a response to mounting concern on the part of the authorities, who saw the connection between the high level of unemployment among young urbans in Johannesburg and Pretoria, the expansion of *tsotsis* (the predominant types of delinquents and criminals in townships), and political unrest in the townships (Evans 1997, pp. 84–90). These policy orientations were enshrined in the law of 1952 to 'stabilise a section of the African urban proletariat', introducing strict differentiation between the category of migrants required to register in labour offices and find work in an urban area within 72 hours and those who were granted urban residential rights, a special provision of the law known as 'Section 10'. The latter included 'those who were born in the city and have lived there permanently; those who have been working for an employer in an urban area for at least 10 years; those who have been legally residing in the urban area for a period of at least 15 years; the wife, unmarried daughter or minor son of the previously defined native' (Hindson 1987, pp. 68–69).

For the first time in South African history, the law of 1952 officially recognised the urban residential rights (instead of mere exemptions) of a category of Africans. While exemptions had formerly been granted on the strength of property, the right to vote, occupational qualifications and services to the state, the residential rights of urbans were henceforth based solely on birth, years of residence or years of employment in the city (Hindson 1987, p. 62). This recognition of rights was premised on the exclusion of all other categories of people from urban areas, especially temporary migrants who were sent back to reserves – henceforth called 'Bantustans' – if they failed to find work. Section 10 laid down the conditions for preferential allocation of jobs, housing and services to holders of residential rights, thereby protecting urbans from the competition of rural workers (Hindson 1987, p. 63). Bureaucratic control intensified significantly: the number of offences related to the absence of valid passes rose from 232,000 per year in 1951 to 414,000 per year in 1959 (Feinstein 2005, p. 155). The situation of migrant workers grew increasingly precarious, even though they continued to be preferred by employers.[7] Illegal migrants had to accept whatever job they were offered by the labour office, unlike urbans who could seek out employment of their own choosing. White employers were not averse to hiring them, but promptly took advantage of their irregular status to pay them less and send back to the reserves those who complained about their working conditions (Pogrund 1990, p. 70). Migrant or 'isolated' women were the other main targets of this new policy. Although in 1952, Verwoerd had to defuse political mobilisation against his plan to extend the pass system to women, the National Party's sizeable re-election victory in 1953 enabled him to 'recommend' that African women be issued a pass in 1954, and make them compulsory in 1960. Eventually, the main difference between the law of 1952 and the law of 1923 lay in the fact that the new law was implemented by the NAD. The NAD had considerably consolidated its power at the expense of the municipalities: it could count on census-taking to provide broader information on urban populations and its jurisdiction could now include an area inside the city limits.[8] The NAD has especially the responsibility to enforce a nationwide registration of all Africans, their names, locale, tax status, fingerprints and their officially prescribed rights to live and work in the towns and cities into a reference book referred to as *Dompas* by Africans or the stupid pass (Breckenridge 2014, p. 138). It was against this backdrop that pass laws were enforced in a particularly brutal manner in the townships of Sharpeville and Langa.

Reinterpreting the Riots in Sharpeville and Langa

Tom Lodge has written the main work on Sharpeville, in which he painstakingly reconstructs the events of this iconic protest. The author pays special attention to the sociology of the PAC leadership and describes the considerable deterioration in the living and working conditions of the Sharpeville residents and Langa migrants. Using sources obtained from PAC activists, Lodge demonstrates that, by recruiting evicted residents and unemployed school-age youths in Sharpeville, the PAC branch in the area was extremely effective in mobilising the local population against the pass policy (Lodge 2011, pp. 79–93), due especially to its task force that urged or even required many residents to join the demonstration. Nevertheless, the analysis leaves unanswered questions about whether or not several key actors participated in the events of both Sharpeville and Langa and the ambivalent effects of the apartheid regime's labour policies on political mobilisation. While the introduction of a compulsory pass for women met with considerable opposition elsewhere, Lodge claims that was not the case in Sharpeville; instead, the women lined up *en masse* to register at the labour office, in particular because the local branch of the ANC had not denounced the new policy (Lodge 2011, p. 82). But then how are we to interpret the main conclusion of the Truth and Reconciliation Commission published in 1998, which supported the testimonials of Sharpeville survivors who stated that most of the participants were unarmed, apolitical women opposed to the pass? (Truth and Reconciliation Commission of South Africa Report 1998, p. 537). Furthermore, why did the young urban 'thugs' known locally as *tsotsis*, who, according to Tom Lodge and the Wessels Commission that conducted the inquiry in the months after the protests, surround the PAC headquarters in Sharpeville demanding the abolition of passes, a demand that did not concern them since, as urbans, they were not legally required to have one?[9] Why did the unmarried migrants in hostels fail to join in the demonstration when they were the first to be concerned by pass inspections, and why, on the contrary, did they mobilise in such large numbers in Langa? While PAC militants obviously played a role in mobilising the people in both townships, it seems necessary to insist on a broader aspect that Lodge overlooked: the sudden application in these two townships of a system that abruptly excluded numerous categories of urban residential rights.

In 1960, the people living in Sharpeville township had been evicted from a slum area called Topville, located in the middle of the region's steel mills of Vereeniging (Chaskalon 1986). In the 1950s, the African population of Vereeniging was housed in the two starkly contrasting townships of Sharpeville and Topville. Sharpeville was a new, tightly and model-controlled township while living conditions in Topville, especially housing were precarious (around 10,000 people housed in shacks) due to a considerable population growth and the reticence of the municipal council to build houses (Chaskalon 1986, p. 9). The African population preferred Topville, however, the rent was not excessive, the steel factories were close to their dwellings, and, above all, the municipal control was limited: there were numerous *shebeens* (illegal taverns serving alcoholic drinks) and police raids failed to eliminate them, thereby allowing many women to earn a living. Until the end of the 1950s, the pressure concerning passes was mild: it was possible to obtain a pass or a residence permit by bribing a member of the Advisory Board.[10] Since 1941, the township of Sharpeville had been slated to replace the shack areas of Topville. Ten years later, Sharpeville was hailed as the best-managed township in the Union and on its way to becoming the prototype of neighbourhoods intended for stabilised urban families. It was endowed with all the modern amenities: six schools, a library, a childcare facility, five football pitches, five tennis courts, electricity in certain neighbourhoods, a stadium, Standard Bank offices and one or two bedroom family homes (Chaskalon 1986, p. 12). These facilities were financed by the rapid growth in sales at the beer hall and the municipal brewery, which had become the third largest in the country by the end of the 1950s. There was little political consciousness in Sharpeville township; the ANC defiance campaign in 1950 had never caught on there, and until 1958, there was no apparent sign of discontent. Nevertheless, there were clear-cut social divisions.

Sharpeville is a good example of the contradictions in apartheid stabilisation policies and the refusal of white employers to comply with governmental directives. Bureaucratic foot-dragging and lack of financing forced the NAD to halt relocation from Topville to Sharpeville in 1951. The NAD refused to enlarge the compounds for workers at the Union Steel Corporation (USCO), the main employer in the area, because they were situated in a white residential area. Their reluctance was also motivated by the official policy of giving job priority to township residents in Sharpeville. Out of a total male workforce of 20 000 living in Vereeniging in 1953 around 11 000 were migrants from Basutoland and the Orange Free state reserves (Chaskalon 1986, p. 7). The employers preferred to continue hiring migrant workers, who were cheaper and more cooperative than the township's school-age youths who showed little inclination for unskilled labour (Chaskalon 1986, pp. 7–9). Furthermore, the municipality played an ambivalent role: on the one hand, it refused to obey the Native Affairs department's orders to send migrants to work on farms to encourage those with rights to remain in the city to take up local urban employment;[11] on the other hand, it failed to convince employers to hire unemployed youth from the township. The number of these youth was very high; they formed groups who spent much of their time mocking migrants who gathered at the municipal beer hall to drink together, engaging in a type of sociability opposed to the consumption habits of the *shebeens,* with which the beer halls were trying to compete (La Hausse 1988). The growing social division between migrants and urban youths was compounded by more violent types of opposition starting in the mid-1950s. The neighbouring town of Evaton was in the grip of factional struggles between a gang known as the Russians, groups of Basotho gangsters who organised after the Second World War to provide paid protection to Basotho migrants working in the region's mines and industries (Kynoch 2005). Following struggles between these two factions in 1956, the police, normally tolerant towards an organisation that had proved useful in combating the ANC or the *tsotsis*, ordered both factions to leave Evaton (Kynoch 2005, pp. 112–114). They found a haven in Sharpeville and Topville, generating renewed violence in 1957 and trouble for the municipality (Chaskalon 1986, p. 20).

In 1958, the fragile equilibrium of the area was disrupted by four new measures. First, the municipality was finally granted an authorisation to remove the 10,000 inhabitants of Topville to Sharpeville, which was accomplished within a year, even though the programme for new housing construction was not yet completed. As a result, much of the evicted population was either assigned to more precarious and costly accommodations or families were crowded together two to a house. About 5,000 Topville residents preferred to move to Lesotho or the nearby town of Evaton until housing conditions improved in Sharpeville (Chaskalson 1986, p. 27). Second, since 1955, the ministry had curtailed the residential rights of migrants from the British protectorate of Basutoland, despite protests from their employers (Chaskalson 1986, p. 21).[12] Consequently, thousands of Basotho who had fled Sharpeville lost the right to return and settle there, and were henceforth separated from their families that had decided to stay in Sharpeville. Third, the construction of the police station in 1958 freed the municipal police from previous constraints and it stepped up its raids in the new township. According to Chaskalson, the chief victims of the increased identity checks were 'the illegal relatives of residents with urban rights' (Chaskalson 1986, p. 22), in other words, the migrant workers mentioned earlier as well as women. When passes became compulsory for women in the neighbourhood in 1958, the women went in large numbers to obtain them. The women's compliance with the measure seems to have stemmed less from the lack of reaction on the part of the local ANC, as Tom Lodge suggests (Lodge 2011, p. 82), than from an urgent need to avoid expulsion to reserves or to Basutoland. The growing frequency of police raids was indeed aimed at actively opposing female proprietors of the *shebeens* competing in Sharpeville as elsewhere with municipal beer halls. The beer halls were the principal source of financing for low-cost family housing construction in the municipality, which had seen its revenues drastically reduced since 1959 following population decline. Fourth, and lastly, the municipality decided to combat the *tsotsis* more vigorously by prohibiting young people from frequenting beer halls and empowering the Advisory Board[13] members to create a group of vigilantes (see Chapter 3) authorised to administer corporal punishment to township youth (Lodge 2011, p. 82). The new population management system already mentioned, which the administration called the 'reference book', strongly identified in the minds of local residents with the pass system, had been in use for several years to clamp down on young people over the age of 16: even very minor offences provided legal grounds for expulsion to the agricultural labour camps that had sprung up since 1955 (see Chapter 2). This administrative practice was backed by the vigilante group, who saw it as a practical way of putting supposedly unruly youths to work (see Chapter 3).

In short, the authoritarian eviction from Topville to Sharpeville and the internal contradictions of apartheid policies had disrupted the neighbourhood's political economy to the detriment of three main groups: women with residential rights but whose incomes dropped even as rental costs rose significantly; school-age urban youths, described by the Wetchel Commission as *tsotsis*, who were in reality denied access to skilled occupations (reserved for whites only) and jobs in industry (reserved for migrants by local employers), and who were being deported on a vast scale to the countryside at the end of the 1950s; and finally, the women and men whose spouses had been refused the right to return to Sharpeville after 1959 because they were from the British protectorate of Basutoland. These were the three principal groups that mobilised against the abolition of passes following the PAC campaign. Not much is known about the unmarried migrants housed in Sharpeville's new hostels, but they did not join the demonstration, unlike the migrants in the Western Cape Province who were the primary target of apartheid stabilisation policies.

In 1955, the municipality of Cape Town signed on to the new preferential labour policy in its local variant, the Coloured Labour Preference Policy, which prohibited hiring Africans as long as there were coloureds seeking jobs in the province. The policy was adopted in response to the demands of coloured voters in municipal elections, who generally supported the measure for fear Africans would compete with them for jobs and housing (Lee 2009, p. 20). At the same time, the municipality and the NAD set about eliminating the 58 squatter areas located on the outskirts of the city, which had exploded during the war – in 1946, the Vindermere squatter area alone may have had as many as 15,000 inhabitants of various origins (Field 2001). In 1954, when the area was rezoned by the NAD to become part of the municipality, the squatters' dwellings were demolished, illegal African population were sent to reserves, and legal Africans, coloureds and white populations were gradually rehoused in the townships and various suburbs. In 1956, the African families that worked legally in town or enjoyed urban residential rights were sent to an emergency camp equipped with only water supply and latrines, where they had to build their own houses; later on the camp became the township of Nyanga West, a typical example of how apartheid stabilisation policy was enacted.[14] The NAD also decided that Langa, the first Cape Town township to have a 'married quarter', would become an area for temporary migrants, particularly 'bachelors', the term used in customary law to designate single men.[15] Finally, in 1956, 9,000 'bachelors' formerly scattered among the various squatter areas on the peninsula, were identified, classified and quartered in 849 'barracks' (worker hostels) that each accommodated 16 men.[16] By 1960, Langa and its extension to Nyanga East had a population of 19,000 men, 18,000 of whom were single[17]. Unlike Sharpeville, where women and school-age youth were most affected by forced removal, in Cape Town the most radical disruption of lifestyles and working conditions took place among these hostels for migrants. From the relative freedom of the squatter areas, where, despite the surrounding poverty, a family atmosphere had developed and a local social life was organised around *shebeens* (Windermere had 900 *shebeens* before the eviction) (Field 2001, p. 33; Lee 2009, p. 26), the migrant workers found themselves in an atmosphere of exclusively male hostels, where they were overcrowded and controlled by a growing number of civil servants and police officers.[18] They also came under tighter surveillance: in Langa, the number of offences resulting from failure to produce a passbook rose from 15 in 1946 to more than 1,500 in 1954 (Muthien 1994, p. 81). Migrants and more general people without the rights conferred by Section 10 were especially targeted: if they loitered in the streets and refused jobs in town, they were turned over to the police, who sent them to farms in the region. The situation had also become critical for African women in Cape Town: since 1954 they were required to have a work permit or risk expulsion (Muthien 1994, p. 69). By 1959, the local Bureau of Native Affairs had already distributed 12,000 reference books, a record number for South Africa (Lee 2009, p. 20; Muthien 1994, p. 70). A delegation of African women complained to the manager of native affairs that Cape Town was the only city in South Africa that systematically applied the Section 10 clauses of the 1952 law to women as a whole, whereas in Kimberley, Section 10 was applied only to 'undesirable' women, i.e. those who owned *shebeens* or who had left their husbands (Lee 2009, p. 20).

Thus, in contrast with Sharpeville, in Langa those who mobilised against the pass laws were neither school-age youths nor very limited number of women, but rather migrant male workers, who had seen their living and working conditions deteriorate in just a few years and their female partners deported to Eastern Cape. In March 1960, Sobukwe, the national leader of the PAC, acknowledged that out of a total of 20,000 party members, 8,000 came from Cape Town, well ahead of Johannesburg (1,000) and Eastern Cape (1,000) (Pogrund 1990, pp. 89, 106, 112, 120–121). Of the 32 party leaders and managers in the province, 10 came from the migrant district of Langa (Lodge 2011, p. 125). In contrast, neither the coloured neighbourhoods were concerned by these new rules nor even the families in Nyanga or those Tom

Lodge calls the 'middle classes' of the townships, joined in solidarity with the Langa migrants during the days of protest (Lodge 2011, p. 125).

The events in Langa, like those in Sharpeville, were not based on class or nationalism; they were partial mobilisations that revealed how fragmented the subaltern groups actually were. They were a protest against the sudden application of a system, which had fundamentally upended the already precarious daily lives of certain groups that had been spared to some extent until then. Lodge's analysis grasp these events in terms of anti-apartheid resistance, which it was of course, but omits to mention the fragmentary character of the struggle that explains his interpretation of the weak response of women to the pass system and the absence of solidarity on the part of the 'middle classes' with 'the people and their struggle'. In so doing, he misses the point: aside from the cases of Langa and Sharpeville, the vast majority of people did not even attempt to resist bureaucratic constraints in the late 1950s. Most migrants and women had no other choice but to comply, including by bribing local civil servants, whereas the so-called middle classes, who for the most part enjoyed residential rights like the migrants with labour contracts in Sharpeville, were not necessarily ready to mobilise in support of the abolition of passes, a system that gave these groups relative advantages in the labour market and in access to housing.

Finally, Tom Lodge claims that the leadership of the demonstration in Langa was made up of 'semi-urbanised' workers (Lodge 2011, p. 126), a term borrowed from a 1963 monograph on Langa by Monica Wilson and Archie Mafeje (Wilson and Mafeje 1963). The two anthropologists, who conducted surveys in South Africa and Southern Africa, were influenced by the Rhodes Livingstone Institute (RLI), founded in 1937 in Northern Rhodesia (now Zambia) and the Manchester School of Anthropology, which saw cities as prime laboratories for social change, modernisation and proletarisation (Ferguson 1999, p. 25). They were trying to understand what they called the forms of 'urban adaptation' manifested by migrants and they sympathised with the demands of African urban residents for access to decent housing, paid labour, schooling and health care, against the wishes of white settlers hostile to any permanent African settlement (Ferguson 1999, pp. 26–37). Their studies were underpinned by modernisation theories and a meta-narrative of the transition as a way for rural dwellers to become modern urban members of industrialised society (Cooper 1983, p. 12; Ferguson 1999, p. 33). But they also demonstrated, that labour and controlled urbanisation policies had far-reaching social effects. Wilson and Mafeje or Mayer in Port Elizabeth discovered that the internal classification criteria adopted in the early 1960s in the township were based on the distinction between the urbanised population on the one hand, and migrant, semi-urbanised workers on the other – a stereotypical distinction used by city residents to describe badly dressed 'country bumpkins' eager to be urbanised (Cooper 1983, pp. 13–31; Mayer 1961). Thus, they were able to show how the actors had re-appropriated the bureaucratic apparatus: urban residential rights were used to guarantee the urbanised population a way of life and social distinction, which they continually invoked to distinguish themselves from those opposed to their lifestyle or who were denied it. Around the same time in South Western Nigeria, anthropologist Abner Cohen, who belonged to the same group of RSI anthropologists as Mayer or Wilson discovered in his empirical study of Ibadan the cultural and political effects of assigning a specific group to reside in a particular urban district reserved exclusively for that group (Cohen 1969). He was the first to expose several decades of colonial management designed to dissociate natives from non-natives, which was essential to Nigeria's social differentiation policies.

Differentiating Natives from Non-Natives in Nigeria

In the early twentieth century, neither the Colonial Office in London nor the local administration in Nigeria was particularly concerned about who had temporary or permanent residential rights in urban areas. Following the conquest of Nigeria, their top priority was to determine what should be done with migrants who did not come under the jurisdiction of native authorities, who were the cornerstone of the architecture of the indirect rule the British sought to establish at the time. The local administration decided to grant certain migrants a separate place of residence (reserved sections of the city called Sabon Gari or Sabo) with specific institutions and rights that differed from those of the local population, thereby creating a lasting distinction based on people's native or non-native origin. Let us go back and take a brief look at these policies using the examples of Kano and Ibadan, before returning to examine the Kano riots in greater detail. Just as the events in Sharpeville help us to understand the substantial divisions created by differentiated management of migrants and urbans, the Kano massacres can also be interpreted as the result – by no means unavoidable – of policies that distinguished between natives and non-natives.

The Kano riots, which lasted from 16 to 19 May 1953, took place in the context of a struggle among Nigeria's leading political parties at the time of decolonisation. The nationalist parties of the south – the Action Group (AG) and the National Council of Nigeria and the Cameroons (NCNC) – wanted to move the date of the country's independence forward. On 31 March 1953, one of the leaders of the AG proposed to set 1956 as the deadline, but Ahmadu Bello, leader of the Northern Peoples' Congress (NPC), the main party of the north, was opposed to the idea, arguing that his region was not ready. The NPC delegation was consequently harassed at the end of the National Assembly session in Lagos. Two months later, the NCNC and the AG organised a meeting in Kano on 16 May 1953 to campaign for independence, which the north opposed. In turn, NPC activists attacked the meeting, setting off four days of violence during which the residents of the Sabon Gari clashed with the other inhabitants of the city. Contrary to Sharpeville, not a single soldier or police officer fired on the rioters (Northern Region of Nigeria 1953, pp. 37–38). The civilian population was entirely responsible for the particularly heavy toll of violence – 36 killed, 336 wounded – and the torching of numerous properties.

Today there are two main interpretations that account for the first large-scale urban riots in colonial Nigeria. The commission of inquiry set up after the riots concluded that 'tribal oppositions' linked to constant criticism of the traditional authorities and political leaders of the north had ignited public opinion in Kano (Northern Region of Nigeria 1953, pp. 39–40). Similarly, the high number of deaths among the Ibgo population from the south-east led John Paden to view the events as the manifestation of relatively spontaneous violence, fuelled by years of economic rivalry between the Hausa and Igbo people, an interpretation supported by many historians and political scientists (Albert 1994, pp. 111–138; Osaghae 1993; Paden 1973). Douglas Anthony, author of the most thorough research to date on the anti-Igbo pogroms in 1966 prior to the Biafran war (1967–1970), offered a different interpretation: he suggested that in 1953 as well as in 1966, the rioting had undoubtedly been instrumentalised for political ends on the first day, but afterwards it took on a life of its own (Anthony 2002, pp. 34–36). Though agreeing on the whole with Anthony's interpretation, I would argue that the Kano riots stemmed neither from nationalist demands nor from ethnic or class-based conflict, but first and foremost from differentiated management of native and non-native populations. When this policy was called into question during the 1940s, it sparked growing hostility between the two groups. It is necessary to trace this antagonism back to its roots, partly forgotten in Nigerian historiography, and re-examine the dynamics of the riots insofar as they revealed the grievances accumulated on both sides and the manifestations of a social order in dispute.

The Birth of Territorial Enclaves: Non-Native Neighbourhoods

In Nigeria, as in most of pre-colonial Africa, there were identifiable differences between host populations and foreign populations. In most West African cities, trading and religious communities (Hausa and Dioula) dispersed across the region had long been living under the control of local authorities: the Hausa were permitted to engage in their activities by authorities, who in turn derived substantial advantages in terms of enrichment, prestige and advice (Schack and Skinner 1978, p. 5). Those relationships were profoundly altered by colonialism, reifying the boundaries separating hosts from foreigners (Lentz 2006, pp. 1–34).

The presence of Christians who had migrated from the south to the north of Nigeria soon became problematic for the colonial authorities. On the one hand, the colonial power was hard-pressed to set limits on its policy of freedom of movement at a time when skilled employees from the south were becoming indispensable to both the administration and Western trading companies (Albert 1996, p. 93). On the other hand, the administration feared unrestricted migration of Christians from the South, who were unlikely to become fully integrated due to their unwillingness to submit to the emir of Kano, the leading native authority of the city and the province. This fear was perhaps exaggerated. Nigeria's northern city-states welcomed and integrated West African and North African immigrant tradesmen, a tradition no doubt strengthened by the establishment of the Sokoto Caliphate in the early nineteenth century.[19] A new section called Fagge, nicknamed the *garin barki* ('visitors' city' in the Hausa language) was created at the city entrance for Kanuri, Touareg and Peul tradesmen, but, according to Paul Lovejoy, this arrangement was above all intended for the type of professions exercised by these tradesmen, whereas wholesale livestock traders had to be located on the outskirts of the city (Barth 1857, p. 463; Lovejoy 1980, p. 53). Indeed, during the same period, tradesmen from the south (Oyo, Ogbomoso and Ilorin) were allocated land and the right to build within the city walls, leading these families to adopt the language, eating habits and religious practices of their hosts (Olaniyi 2005a, pp. 88–92).

In 1909, the new colonial administrator, the 'Resident' of Kano district, Charles Temple, who was proficient in indirect rule, asked all the Nigerians from the south and other West Africans who had formerly been scattered throughout the city to relocate to the military quarters set up in 1904. In 1911, the area acquired its definitive name: Sabon Gari ('new city' in the Hausa language), which was located one kilometre outside the Kano city wall. The main reason for establishing the Sabon Gari was to create a neighbourhood for 'native foreigners' who were not under the jurisdiction of the native authorities. The Sabon Gari thus became an administrative enclave run by a British station magistrate answering directly to the Resident, unlike the rest of the city, which remained under the authority of the Emir (Fika 1978,

p. 211). Lebanese and European companies and traders from the south were barred from settling inside the walls of Kano. In 1913, the Emir expelled all non-Muslim tradesmen from the old city and prohibited the sale of land and houses in that section to outsiders (Cristelow 2005, pp. 256–262; Ubah 1985, p. 93). Yoruba Muslims were allowed to live in the old city provided they complied with Muslim practices (no alcohol, veiled women) (Barkindo 1983, p. 14; Olaniyi 2013, pp. 67–89).

During the 1910s, a similar process was under way in other Nigerian cities: neighbourhoods were reserved for Nigerians from the south in the old cities of Gusau and Zaria and in the new towns of Kaduna and Jos in Northern Nigeria, and neighbourhoods were reserved in Lagos, Ibadan and Bamenda, where Hausa tradesmen were required to live after the eastern part of Cameroon was attached to Nigeria in 1918 (Awason 2003, p. 292; Home 1997, p. 130; Paden 1973, pp. 113–114; Plotnicov 1967, p. 41;). The administration of a specific quarter reserved to northern Muslim immigrants in the city of Kumasi (Gold Coast) 'was also modelled after that of the Hausa emirates in Northern Nigeria' (Bigon 2016, p. 212). All these initiatives were linked to a broader policy pursued by Lord Lugard, the Governor General of Nigeria from 1914 to 1919, who recommended that the townships be reserved for native foreigners, whereas local natives, even those employed as simple labourers by European employers, were to remain under the control of the native authorities (Kirk Greene 1968, p. 163). The law of 1917 harmonised these local arrangements by defining the township as an enclave outside the jurisdiction of native authorities, which 'consequently relieved these authorities from the difficult task of controlling native foreigners, government employees and Europeans' (Home 1997, p. 128). The law recommended establishing enclaves for Nigerians from the south who were living in the north to be called 'Sabon Gari', and enclaves for the populations from the north living in cities of the south that would take the name 'Sabo', a diminutive of Sabon Gari (Oyesiku 1998, p. 41).

Residential segregation along with parallel institutions to govern these differentiated populations continued to be the dominant model in the south until the 1940s and lasted until independence in the north. In Ibadan, as Abner Cohen demonstrated, the interests of the colonial administration coincided with those of the chief of the Hausa community in Ibadan (the Sarkin Hausa). Indeed, this chief possessed sufficiently broad powers over the administration, justice and the police[20] to group all of the city's Hausa tradesmen together, and thereby keep control over the regional market of cola and livestock in a context of growing competition with local populations. Sabo maintained its ethnic exclusivity throughout the colonial period,[21] even though it could no longer accommodate all the residents of the north after the 1940s.[22] The Sabo chief nevertheless remained under the authority of the native chief of the city named Olubadan, who answered directly to the Resident (Cohen 1969, p. 162).

In the cities of the north, on the other hand, the Sabon Gari were administrative enclaves controlled by district British administrators until 1940, when they were replaced by the authority of the emirs, setting off fierce protests from the inhabitants (see below). In Kano, starting in 1932, the administrator was assisted by an advisory board comprising a president and two representatives of the Sabon Gari communities who helped him collect taxes and oversee relations between the Sabon Gari, the emir of Kano's advisory board and the colonial authorities.[23] The district had its own court of justice; the only police force that could intervene was the Nigeria Police Force (NPF), contrary to Kano, which was under the supervision of Native Authority Police. Alcohol and beer could be sold in the Sabon Gari communities (whereas it was strictly forbidden in the city), women were allowed to live alone, and schools could be set up by missionaries or financed by local associations (whereas the administration was hostile to the introduction of Western education in Northern Nigeria until 1945) (Olukoju 1991, p. 363; Tibenderana 1983). The implementation of these policies does not seem to have caused any major conflicts until regionalism developed after the Second World War.

Regionalism and Decolonisation

The regionalisation process in the 1940s and 1950s involved transferring the functions of the colonial administration to the three main regions of Nigeria (east, west and north). It was accompanied by a new policy of regional self-identification, which Eghosa Osaghae called 'regionalism': 'a system in which citizens who are not natives of a given region are discriminated against and excluded from access to public goods. In other words, the government of the region gives access to public goods only to citizens who are natives of that region' (Osaghae 1998, p. 7). The northern region championed regionalism (Osaghae 1998, p. 9): in 1952, in an effort to close the gap in school and university training following indirect rule, Ahmadu Bello, leader of the northern region, launched a plan to train, employ and promote the people of the north (now described as Northerners) in the civil service, a process called Northernisation, which became the priority of his government during the 1950s and 1960s (Paden 1986, pp. 252–257). At the same time, Ahmadu Bello sought to limit the access to civil service jobs, land and public procurement contracts to people of the south (henceforth called Southerners) (Anthony 2002, p. 44; Paden 1973, p. 319; Paden 1986, p. 256). Tellingly, the terms 'Southerners' and 'Northerners' tended to replace the terms natives and non-natives, and became their equivalents in the jargon of administrators and politicians.

At the end of the 1940s, the formation of regional political parties (the NPC, the dominant party in the north, the AG, the dominant party in the south-west, and the NCNC, the dominant party in the south-east) was indeed the springboard for manufacturing the regional antagonisms familiar to historians of Nigeria. First of all, the opposition between the north and south: the parties of the south were in favour of independence, whereas the leaders in the north were against it, fearing that the lack of qualified administrative personnel from the north would lead to domination of the state and the regions by the elites of the South. Second, the conflict amongst the so-called majority ethnic groups (Yoruba, Hausa and Ibgo) championed by these parties for control over national and regional leadership. Third, the increasing divisions between the majority groups and minorities in the same regions, which then formed opposition groups and allied themselves with the dominant party in the rival regions. It should be noted that the term 'ethnic minority' did not enter the vocabulary of the political elites until the regionalism process was under way (Osaghae 1991, p. 238). Regional political oppositions were exacerbated in Kano, where the NPC was dominant in the old city, whereas a minority radical party (the Northern Elements Progressive Union (NEPU)), still allied to the NCNC, dominated not only in the Sabon Gari, but also in numerous other neighbourhoods of the city, which wanted to end the all-powerful hold of the emir and his administration. In the case of Kano, the regional antagonisms were amplified by a conflict linked to the management of local affairs in the Sabon Gari starting in 1940.

Cooperation between the Sabon Gari representative and the native authorities deteriorated sharply in 1940 when, due to wartime restrictions, the colonial administration decided to put the Sabon Gari enclave back under the authority of the emir.[24] According to John Paden, this retreat created more tension between the local and 'foreign' populations (Paden 1973, p. 335). The civic and ethnic associations, created during the war to promote and defend the special interests of the residents, forcefully objected to the measure. They demanded a local democracy (i.e. a town council with elected representatives to replace the emir's appointees) and the elimination of separate institutions for managing Kano affairs (i.e. the abolition of the township council for Europeans and the Advisory Board for Africans), and denounced the emir's new interventionism in their local affairs.[25] A township representative summed up these fears in 1941:

> The District Officer of Kano division said that we, Southerners, are only here in Kano *don albarkarchin Sarki* ('on the goodwill of the Emir').

Hausa men live in Egbaland in peace why shouldn't we live here? If we are law abiding, why should we be here 'on the goodwill of the Emir' and not of right as Nigerians? We object to the idea of permission. We are here as Nigerians – of right anywhere in Nigeria.[26]

The dispute focused primarily on the rights of Sabon Gari inhabitants to residence and 'ownership'. The issue of housing was becoming a concern: the district population had tripled between 1939 and 1954 from 8,000 to 21,000 inhabitants, whereas it was designed to accommodate about 10,000 This rapid growth was mainly due to the increase in the Igbo population, which made up two-thirds of the inhabitants in the mid-1950s (Paden 1973, p. 258). Starting in the 1930s, medical surveys noted the squalor in the district, which favoured the business of slumlords reported by the doctors: 'The plot holders act as they will, building dark, poorly ventilated rooms to let, and the tenants pay considerable sums for plots where 35–40 people are crammed together, not counting children'.[27] All the surveys taken during this period end with the same recommendations: to reduce the land-to-building ratio for construction, prohibit the accumulation of plots and their sale without notifying the administration and improve existing housing to prevent epidemics.[28] Although the district's unsanitary conditions were a long-standing concern, the new regulations on neighbourhood buildings did not come into force until 1940, when the Sabon Gari were transferred to the native authorities, who were given significant policing powers.

The initial preoccupation with sanitation was soon eclipsed by a far more vital issue: the very terms governing the settlement of Southerners in the northern region. The transfer of power to the native authorities had especially increased the control over Sabon Gari residents, particularly by the Native Authorities Police (known as the NA police or 'Yan Gadi' in Hausa) that had replaced the NPF in 1940 (Chiranchi 2001). That same year, the Sabon Gari was divided into sections, each one under the authority of a chief who had a map of the various buildings,[29] along with detailed information on their occupants, the identity of the leaseholder, his business activities, and the rent he charged his tenants. Every week a member of the NA police visited the section chief to register the arrival of new immigrants and verify any violations of local regulations. This was a major change. As Bola Ige points out, whereas residence certificates were formerly delivered by colonial administrators, that role was now assigned to local native civil servants (Ige 1995, p. 99). A group called the Sabon Gari Plot Holders Association was organised to protest against this supervision and the new forms of perceived discrimination:[30] plot holders could no longer sell their homes to the highest bidder or to a person of their choice without first informing the native authorities, who could prohibit the sale and propose a different buyer; they could no longer mortgage their homes to take out loans, nor own more than one plot per family; women, whether married or not, could no longer hold leases; leases were renewed at the discretion of the local authorities, no matter how much the plot had been developed.[31] These rules applied only to residents of the Sabon Gari and not to residents of the European neighbourhood. In 1949, the association demanded 'an end to these kinds of discrimination before they are turned into a riotous demand by agitators'.[32]

The Kano Riots

The residents of Sabon Gari thus found themselves in a subaltern position on two fronts: in relation to the colonial administration that employed them and in relation to the native administration that henceforth determined how they were to live in their neighbourhood. They considered themselves victims of twofold discrimination: the first was colonial in nature, imposed by the British administration, which continued to refuse the residents the same political and social rights as those of Europeans (hence the demand for a unified town council); the second was a new and far more pervasive form of discrimination, emanating from the native administration, which sought to limit the rights previously granted to residents during the colonial period. For the protagonists of the Sabon Gari, the riots were the manifestation of built-up resentment towards the new native administration, a revolt against the new political configuration. To the residents of the old city, the attacks on the minority population in the Sabon Gari were an extreme sign of an increasingly discriminatory ideology directed against Southerners and promoted by the native authorities and NPC leaders and militants. In this context, the actors in both camps used violence, rumour and looting to disrupt a social order they perceived as unfavourable to them.

The newspaper *West African Pilot*, founded in 1937 by Nnamdi Azikiwe, leader of the NCNC, compiled the commission of inquiry's conclusions based on interviews conducted three months later in Kano, reporting that youths described as hooligans were the first to attack the residents of the Sabon Gari on the evening of the 26 May 1953 meeting campaigning for independence organised by the AG and the NCNC, and again the following day in the hope that 'the disruption would bring more opportunities for looting'. In the 1940s and 1950s, it was common practice for political parties to recruit thugs (called 'area boys' or *'jaguda* boys' in the South, and *'yandaba'* in the north) to serve as enforcers (Anifowose 1982, pp. 230–234; Rotimi 2001, p. 140; Ya'u 2000). In the case of Kano, there is no direct proof of such practices, but there is incriminating evidence. The delegates of the AG and the NCNC political parties were planning to organise a meeting on

17 May in the neighbouring town of Kaduna. On 15 May, the delegation leaders received telegrams warning of possible attacks against the delegates, a sign that premeditation cannot be ruled out.[33] The speech delivered the night before the riot by Inuwa Wada, secretary of the Kano branch of the NPC, calling for violence to avenge the affronts committed against the NPC in Lagos two months earlier, also reveals how the NPC stoked existing tensions by using thugs:

> Having abused us in the South these very Southerners have decided to come over to the North to abuse us, but we are determined to retaliate the treatment given us in the South. We have therefore organised about 1,000 men ready in the City to meet force with force.[34]

The partisan political dimension should not be minimised. On the contrary, it was a key factor both in the old city and in the Sabon Gari when ethnic associations became AG and NCNC recruiting grounds at the end of the 1940s.[35] In all likelihood, the violence was less spontaneous than the initial research suggests, and local political leaders, particularly those of the NPC, no doubt played a decisive role in triggering the riots.

The political violence developed in a context of steadily mounting mutual opposition and distrust between the Sabon Gari and native populations, with rumours helping to escalate the tension and violence. Prior to the riot, talk of organised armed groups ready to take over adjacent neighbourhood circulated on both sides, prompting the communities to try and arm themselves. The day before the outbreak on 19 May, the inhabitants of the Sabon Gari bought locally made firearms (referred as *dane* guns) 'knowing that something was going to happen'.[36] They were used not only to defend their neighbourhoods, but also to attack the inhabitants of other areas in the city. At the end of the first day, new rumours led to a spate of renewed violence:

> During the night, stories were circulating in the City and in Fagge, some true, most either false or exaggerated, of retaliatory acts by Southerners in the Sabon Gari, some of which were alleged to have concerned Hausa women... It was not merely a matter of revenge for alleged acts against their fellow countrymen. It was becoming even more a demand for 'preventive war'.[37]

The relationship between the population and law enforcement authorities was another critical component in the production of violence, inseparable from the economic aspect of looting. The entire NA police of Kano (350 men), two NPF anti-riot units (100 men), and a brigade of British gendarmes (130 men) were mobilised to quell the riots (Northern Region of Nigeria 1953, pp. 12, 12–19, 25–26). The NA police were deployed in the Sabon Gari, but in the early morning, the residents attacked the police officers, who were suspected of allowing looting to take place.[38] There is some basis for these allegations. During the four days of rioting, the NA police arrested 120 people only, 62 of whom received sentences of one to four months in prison (Olaniyi 2005b). In contrast, when one small NPF unit (53 men) regained control over security in the Sabon Gari between 24 May and 30 September, 3,232 individuals were convicted and sentenced, and the equivalent of 21,258 pounds in stolen goods were recovered (Olaniyi 2005b). For the NPC militants, looting the goods of Sabon Gari residents and setting fire to their homes merely carried to extremes the discourse of the NPC government and the agents of the native authorities who regularly reminded them of the precarious nature of the Southerners' 'occupation' and rights in the north (Ige 1995, p. 100). For the Sabon Gari residents, on the other hand, targeting the members of the NA police gave them a chance to take their anger out on those they held responsible for the discrimination they suffered and intimidate those perceived as intelligence agents who could inform people outside about the goods waiting to be looted in the neighbourhood.

At first, no institution was able to control the rumours or check the ensuing cycles of violence, and ultimately, none of them had enough legitimacy to impose a ceasefire on the belligerents. The native authorities were the only institution common to the opposing parties, but for years they had been deaf to the numerous demands of Sabon Gari leaders and they were seen not only as illegitimate but indeed as supporters in the hands of the NPC. The failure of the indirect rule to create a space for mediation made way for the central role played by rumour in aggravating tensions and the cycle of violence. Probably the only organisation that transcended the territorial divisions was the NEPU party, which might explain why its supporters, whether Hausa or not, either helped the Southerners to escape from their assailants or refused to take part in the riots and looting (Sklar [1963] 2004, p. 131; Olaniyi 2003, p. 229). In the end, segregated space, the introduction of separate institutions under indirect rule and the reifying of social and cultural differences between 'migrants' or residents of the Sabon Gari and residents of the old city clearly fuelled the violence of the massacres.

Conclusion

To accommodate migrants needed for the colonial economy without calling into question racial hierarchies in South Africa and the native authorities in Nigeria, bureaucrats invented *apparatuses* of exclusion aimed at not granting migrants the same rights nor the same urban space as to the other urban dwellers. Despite the differences of labour policies between Nigeria and South Africa, the classification of migrants as a specific bureaucratic category and their compulsory residence in specific enclaves share four common features that draw the contours of what might be referred to as urban colonialism. If colonialism is the domination of colonised majority by a European minority depriving the former of their fundamental rights, urban colonialism is one of its specific forms that grants different rights (especially in terms of housing and labour) to different colonised categories living or residing in towns and cities.

Firstly, the colonial urban governmentality was limited and did not serve the well-being of the population but provided a legal and informational basis for the discrimination of migrants and their distinctions from the rest of the urban African population. If all the African population was discriminated by colonial and apartheid policies, the migrants were even more affected in their daily lives. During the 1950s, the bureaucracies of the two countries had individualised information through instruments such as reference books, labour bureaux and municipal censuses in South Africa, and surveys on sanitary conditions of neighbourhoods, lists of plot holders and owners, zoning maps and residence certificates in Nigerian Sabon Gari. While the information was incomplete – in Nigeria it pertained to only a few parts of urban space, and in South Africa, keeping files on the population proved to be a failure and a chaotic process for Breckenridge (2014, p. 137, 161) – it nevertheless appeared to be a tool of individual control and individual discrimination against certain categories of the population, which, like those in Kano and Sharpeville, tried to oppose it but very often had no choice but to adjust to it.

Secondly, this form of colonialism found its origin in the urban fabric as cities became laboratories for testing and implementing public policies that were highly exclusive towards migrants. These massacres reveal the particular stories of the townships of Sharpeville, Langa and Kano, where the processes of differentiation between urbans and migrants, natives and non-natives, had been especially exacerbated in the course of the previous decade. They reveal a struggle of subaltern groups (women migrants and tsotis in Sharpeville, migrants in Langa, non-natives, migrants, residents and political activists in Sabon Gari in Kano), opposed to new government measures (mandatory passes in South Africa, new supervision by the Northern Native Authorities in Nigeria and their regionalist policy). These events had the consequential effects of strengthening exclusion at the national level. Sharpeville was followed by the ban on anti-apartheid organisations (ANC, PAC after the South African Communist Party (SACP) in 1950), which went underground and abandoned non-violence.[39] South Africa entered its more repressive period in history. Control over migrants was strongly consolidated from the 1960s to the end of apartheid: the members of the government in favour of opening up urban residential rights were marginalised within the government, the number of people punished for failing to present a pass rose from 414,000 in 1959 to 694,000 in 1967, and the distinctions between migrants and urbans (urbanised, semi-urbanised, migrants) grew sharper in the following decades (Feinstein 2005, p. 155; Posel 1991, pp. 240–241). After Sharpeville, South Africa and the pass system in particular became metonymic for white supremacy everywhere (Breckenridge 2014, 139). The pass laws were officially abandoned in 1986 but regulations for the millions of migrants staying in hostels remained the same: the hostels were reserved for males with *bona fide* employment who were legally allowed to be in urban areas until the end of apartheid (Ramphele 1993). In Nigeria, the Kano massacres led the Colonial Office to convene a conference in London with the representatives of Nigeria's main parties, who negotiated a federal constitution giving greater leeway to the leaders of the various regions to de-escalate existing tensions (Lynn 2006, pp. 245–261). Nigeria became a federal country at this very moment. The strengthening of the executive powers of the regions gave, however, the leaders an opportunity to intensify their regionalist tendency to discriminate against Nigerian migrants from other regions in granting access to public goods, to state employment, to university and to political positions, an issue that has resurfaced again and again ever since (see Chapter 6).

Thirdly, the different rights provided to different categories have been appropriated by the populations and in the case of Nigeria by new state institutions at independence. Within a few decades, the terms natives, non-natives, migrants and urban became part of everyday life. Initially devised to govern populations they became ordinary categories commonly used to designate oneself and others. This appropriation of invented colonial categories had enduring legacy in both countries. In Nigeria, the terms native and non-native were replaced in the 1970s and 1980s by the designations 'indigene' and 'non-indigene' which became common categories used by the administration, the politicians and ordinary people in a new urban politics of exclusion (Chapter 6). In South Africa, the distinction between migrants and urban dwellers having township rights has enduring effects even after the end of apartheid. During the 2008 violence, old residents in the Alexandra township of Johannesburg insisted on their historical rights to housing obtained during apartheid and criticised newly arrived migrants, both nationals and internationals for not wanting to 'join the queue' on the waiting list for future housing (Nieftagodien 2011). In the squatter area of Atteridgeville on the outskirts of Pretoria, the leaders of the squatters' movement who had been struggling since the mid-1980s to benefit from decent housing sent a clear message to newcomers amid the 2008 violence: everyone but especially newcomers (i.e. migrants) had to abide by the 'policy of patience' and the waiting list (Monson 2015). In the lower middle-class neighbourhood of Luloyloville in Cape Town, a number of owners looked down at neighbouring hostels and squatter camps and also considered they have historical rights of decent houses (Buire 2019, pp. 252–255). In other words, the right attached to house and labour invented by the apartheid government to township dwellers is still very much part of the township residents imagination today.

Fourthly, assigning distinct areas to groups, granting them specific rights in terms of housing and work, placing them under a distinct authority, provides an important basis for triggering collective violence. This is probably one of the most important legacies of urban colonialism. Indeed, the new social and spatial realities inherited from decades of differentiated population management were at the heart of local clashes and wider national conflicts. The 1966 anti-Igbo pogroms in the northern region that led to the civil war (1967–1970) were similar but on a larger scale than the events in Kano in 1953 and primarily affected Sabon Gari residents in northern cities, whereas the Southerners integrated in local Islamic communities were often spared (Anthony 2002). Many more localised conflicts since the colonial period (in Warri (Delta State) since the 1950s, in the cities of Jos and Yelwa (Plateau State) since 2001, in Ile-Ife (Osun State) throughout the twentieth century, in Zangon Kataf (Kaduna State) in 1992, and in the metropolis of Kano (Kano State) in 1953, 1966, 1991, and 1996) have rightly been qualified as indigene-settler conflicts (Adebanwi 2009; Ekeh 2007; Higazi 2007; Madueke 2018; Ukiwo 2006). Beyond their particular local histories, these conflicts are all controversies over how to determine who is a genuine *indigene* (or in colonial parlance a native) of the city (HRW 2006, pp. 43–44 and 54–58). Some of these conflicts are a direct legacy of this colonial policy of belonging, others are a re-enactment under new political and economic conditions of the opposition between natives (indigenes) and non-natives (non-indigenes) (see Chapter 6). In South Africa, the recurrent clashes between Marashea (gangs protecting migrants) and *tsotsis* from the 1950s to the 1970s demonstrate that these divisions were profoundly linked to residence (urban vs. rural) and age group (adults vs. youths) (Kynoch 2005). The anti-delinquency committees set up before and after the riots in Sharpeville and Langa attest to both an effort to combat the supposed unruliness of young people and migrants and a developing sense of respectability and civic pride among organisations of parents and elders in the townships (see Chapter 3).

In the end, 'detribalised', 'delinquent' and 'unruly' youths, who, according to the Sharpeville and Kano commissions of inquiry, were quick to convert to political causes, were seen as a major threat to the stability and political order in South Africa and Nigeria. Due to the absence of alternative sources, it is difficult to assess the involvement of young people in these particular events in greater depth. Even so, this reading undoubtedly conveys a deeper political concern shared by administrations, governments and many local organisations. This image of urban youth as marginalised and ready to take part in violence was in turn the product of a very particular history in which we can see how that image might have been – and was in fact – constructed as a danger by a group of actors bent on reforming their behaviour.

Notes

1. South Africa became a dominion of the British Empire in 1910, comprising the former British colony of Natal and Cape colony and the former Boer republics of Transvaal and Orange. The country was endowed with a government and a Parliament elected by the white minority. Its 6 million inhabitants were classified by 'race': blacks (67%), whites (21%), coloured (9%) and Indians (2.5%). Nigeria was a colony of the British Empire, set up in 1914 following the merger of the Northern Nigerian Protectorate, the Southern Nigerian Protectorate and the colony of Lagos. At the time, it had 16 million inhabitants, including 3,000 Europeans residents.

2. See also the resurgence of autochthonous claims in Douala, Cameroon, in Geschiere (2009).

3. The discovery of significant deposits of diamond in Kimberley (1867) and gold in Johannesburg (1886) gave rise to a mining and industrial economy requiring a large workforce from the entire country as well as several colonies in Southern Africa (Mozambique, Lesotho, Swaziland, Nyasaland, Southern Rhodesia, etc.), India, China and Europe.

4. Labour policies have been analysed in numerous historical works in South Africa that have been briefly summarised here, whereas the policies regarding the differentiation between natives and non-natives in Nigeria are much less familiar, and based on documents found at the national archives in Ibadan and Kaduna. The story of Sharpeville is primarily taken from an unpublished monograph by Matthew Chaskalon entitled *The Road to Sharpeville*, African studies seminar paper, University of Wits, African Studies Institute, 1986. The story of Langa relies on the existing literature, supplemented by the archives of

the Advisory Board of the townships of Langa and Nyanga West obtained at the Cape Town archives. The story of the Sabon Gari* in Kano is written thanks to reports available at the National archives at Kaduna. The reports drawn up by the commission of inquiry in Sharpeville in March 1960 and the commission of inquiry in Kano in May 1953, as well as those written after the riot by the journalists in the *West African Pilot* give additional information. Nevertheless, this documentation is insufficient to reconstruct the events with exactitude.

5. The African urban population rose from 600,000 inhabitants in 1904 to 1.7 million in 1936, thus exceeding the number of whites residing in the city (1.3 million).

6. There was a shortage of 154,185 houses for families and a shortage of housing for 106,877 workers in 1947 (Hindson 1987, p. 56).

7. Family housing in the compounds* of mining companies was limited to 3% of the workforce during the 1950s (Crush et al. 1991, p. 13).

8. There is disagreement among the specialists on this point. Contrary to Ivan Evans, Keith Breckenridge maintains that the new centralised system was just as inefficient as the previous decentralised one. The introduction of the reference book created a bureaucratic nightmare for the Ministry of Native Affairs, and the plan to ensure that all South Africans had an identity document with fingerprints had to be scuttled, leading to trafficking in reference books. Breckenridge's article stops at the year 1960, however, and does not talk about the conditions under which the reference book was implemented after that date (Breckenridge, 2005, pp. 102–104). It should be noted that bureaucratic control intensified significantly during the 1950s: the number of offences related to the absence of valid passes rose from 232,000 per year in 1951 to 414,000 per year in 1959 (Feinstein 2005, p. 155).

9. P.J. Wessels, *Report of the commission appointed to investigate and report on the occurrences in the district of Vereeniging (namely, at Sharpeville location and Evaton) and Vanderbijlpark, province of the Transvaal on 21 March 1960*, 23 September 1960, Supreme Court, Pietermarizburg.

10. Hundreds of residents were thus without a pass when they were evicted in 1958 (Chaskalon 1986, pp. 8–11).

11. To avoid alienating local employers, the municipality refused to introduce police escorts to force migrant 'unemployed to find work at farms in the region.

12. In 1955, an amendment to the Urban Areas Act changed the status of migrants from the British protectorate of Basutoland. Henceforth, they were authorised to remain in an urban area only if they had been

legally present on 6 May 1955. They were allowed to change jobs within the urban area, but lost that right if they left the zone for any reason other than holiday leave.

13. Advisory board is a body of co-opted or elected African members who advised the municipal superintendents or managers in their day-to-day administration of African townships and locations. They served as a link between urban Africans and local authorities and they provided a means of social control (Baines 1994, 81). Most of the time managers developed a highly personalised nature in the administration of townships and consequently played an important role in shaping the form of cities and city politics (Musemwa 1996; Robinson 1991; Sapire 1994).

14. In 1960, family housing in the township could accommodate 2,789 men, 3,009 women and 4,313 children.

15. This decision was reported in a number of newspapers: 'New threat to Langa', *New Age*, 9 December 1954; 'Verwoerd denies Langa threat. Statement on Obstacles to city's plans', *Cape Times*, 29 April 1954; City Native Debated. Backing for state in Council, *Cape Times*, 28 January 1955.

16. CTA, 2/OBS 3/1 680, progress report on action taken by the council for the provision of Native housing for the period May to October 1956.

17. CTA, AWC, 3/1, City of Cape Town, town clerk department, native authorities branch. Memorandum in the urban area of the city of Cape Town, undated, about 1960.

18. The number of civil servants rose from 221 in 1947 to 343 in 1958. Yvonne Muthien, *State and Resistance in South Africa, 1939–1965*, Aldershot, Avebury Publisher, 1994, p. 81.

19. In Katsina, Hausa and non-Hausa tradesmen lived in the same neighbourhood; in Kano, North African tradesmen mixed with the local population in the Dalla district (Lovejoy 1980, p. 52).

20. He ruled on civil law cases (marriage, divorce, estate, child custody), and had the power to arrest suspects and transfer them to the police; he also performed the functions of an administrator (ensuring the upkeep of mosques and cemeteries, helping the needy, supervising short-time hotels).

21. In 1916, there were 400 Hausa in the district; by 1963, there were 4,184. During the 1940s and 1950s, when the Yoruba were authorised to own property in the district, only the Hausa were authorised to hire out rooms.

22. National Archives Ibadan, NAI, Oyo Prof 1, 592, Letter from the Divisional Office Land Section, Ibadan, to the Secretary Western Provinces, Ibadan, 12 December 1950.

23. National Archives Kaduna, NAK, Kano Prof, 6115, Organisation of Sabon Gari Administration by Wesport, 1938.

24. The city was divided into three sections: the GRA reserved for Europeans, the pre-colonial city under the exclusive authority of the emir, and finally 'Wage', which included the colonial housing developments reserved for Africans that were non-natives of the city (Fagge, Sabon Gari, Tudun Wada and Gwarwarga, inhabited by migrants from the North) under the authority of the emir, who in turn was counselled by Advisory Boards made up of representatives from the various districts appointed by the emir.

25. NAK, Kano Prof 6115, Representatives of tribal and federated unions of the Sabon Gari to the Senior Resident, Kano Province, 1 November 1944.

26. NAK, Kano Prof 6115, Prcis of talks at a meeting between representatives of the Sabon Gari community and the Resident, 16 November 1941.

27. NAK, Kano Prof, 6122, Medical Officer of Health to Local Authority, Kano, 16 March 1942.

28. NAK, Kano Prof 6122, Medical Officer of Health to the Resident Kano Province, 25 February 1939.

29. At the end of the 1930s, a building plan for each plot was produced by the mapping section of the Native Authorities. NAK, Kano Prof, 6122, Resident Kano Province to the Secretary Northern Provinces, Kano Township Overcrowding in Sabon Gari, 29 March 1939.

30. NAK, Kano Prof 6122; Petition by Kano Sabon Gari Plot Holders Association, 31 March 1945 to the Secretary Northern Province Kaduna.

31. NAK, Kano Prof, 6122, Kano Sabongari Plot Holders Association to the Resident Kano Province, 15 August 1949.

32. NAK, Kano Prof, 6122, Kano Sabongari Plot Holders Association to the Resident Kano Province, 15 August 1949.

33. 'Kano Riots. 'The Wide Scene', *West African Pilot*, August 17, 1953.

34. 'Kano Riots. 'The Wide Scene', *West African Pilot*, August 17, 1953.

35. Thus, the Yoruba Central Welfare Association (Egbe Omo Oduduwa) set up in Kano in 1942 preceded the branch founded by Awolowo in London in 1945 and in Lagos or Ile-Ife between 1945 and 1948, even

before the creation of the Action Group in 1949. The Igbo Community Association, which was a member of the Plot Holders' Association, became a recruiting ground for the NCNC in Kano in 1949.

36. 'Kano Riots. General Conclusions'. *West African Pilot*, August 19, 1953.

37. 'Kano Riots. General Conclusions'. *West African Pilot*, August 19, 1953.

38. 'Kano Riots. General Conclusions'. *West African Pilot*, August 19, 1953.

39. On the respective roles played by the Sharpeville events, the international context (support from the USSR and China) and the internal evolution of the leadership of SACP, the ANC and Mandela in the transition to armed action, see Stephen Ellis (2011).

CHAPTER 2

The Making of a Delinquent

Within a few short years, South Africa (in 1937) and Nigeria (in 1943) both adopted legislation on children and youth, inspired in large part by the criminal reform movement in the United Kingdom, which held that young people were less responsible for their actions than adults, and therefore should not be subject to the same laws. The movement campaigned in favour of raising the age of minors, modifying the judicial process to 'rehabilitate' rather than punish children and youth, and creating a body of magistrates and experts (commissioners of child welfare, juvenile court judges, welfare officers and probation officers) to remove children from a prison environment deemed harmful for them and for their subsequent socialisation. The simultaneous expansion of an administrative and judicial system specialised in providing for the well-being of children 'in need of care' and of the number of juvenile delinquents led to increased state intrusion in the daily life of families, the production of individualised knowledge concerning these target groups, and the introduction of schemes designed to reform their behaviour.

This history, familiar to historians and sociologists of children and youth, is still not well known with regard to Africa. In the past twenty years or so, a number of historical, sociological and anthropological studies have documented the existence of so-called delinquent youths, who live off extortion and violence, and occasionally join political parties or engage in warfare (Veit, Barolsky and Pillay 2011). Most of the research in Africa is focused on the multiple origins of delinquency and street children linked to a particular age group and how its members transgress laws and prevailing norms (Glaser 2000; Jensen 2008; Morelle 2007; Rodgers 2019). While it is difficult to know whether delinquency is a particular kind of social behaviour, it is first and foremost criminal behaviour, i.e. a violation of the law, which threatens its author with punishment (Robert 2005, p. 11). We have chosen here to

Classify, Exclude, Police: Urban Lives in South Africa and Nigeria, English Language First Edition. Laurent Fourchard.
© 2021 John Wiley & Sons Ltd. Published 2021 by John Wiley & Sons Ltd.

adopt a sociological approach to incrimination, which remains unexplored in Africa today even more than in Europe (Robert 2005). This approach accounts for the simultaneous introduction of rehabilitation schemes *and* the use of coercive methods in dealing with children and youth identified as dangerous, in need of care or undesirable, and allows us to question the ambivalent attitudes of the institutions responsible for these groups. This is closed to Foucauldian criminologists who 'draw attention to the impact of new knowledge and technologies upon the power relations between governmental actors as well as between the rulers and the ruled' (Garland 1997, p. 188), in our case between delinquents and a set of welfare and administrative and judicial officers.

This chapter explores the introduction of new penal reform policies, the identification of young people wandering in the street as part of a more general political problem and the setting up of new apparatuses to deal with this new problem. It questions three major theses that have been prevalent to understand the socialisation of youth in cities. The detribalisation thesis dominant during the colonial period estimated that delinquency stemmed from unchecked urbanisation, which broke families apart, weakened the social control of elders over young people or parents over children, offered only unstable job opportunities to young, unskilled workers and facilitated their violent socialisation in the street. In this chapter, we suggest the opposite. More repressive policies concerning urban youth came into being because colonial administrations widely subscribed to the idea that urbanisation brought with it the dislocation of African families. The analysis of the actions of social services and ministries in Nigeria and South Africa must be deciphered in light of this colonial obsession.

The proletarianisation thesis dominant in the 1970s and the 1980s reverted colonial officials' argument. Radical historians tended to view young delinquents and gangs in general as a workforce that was difficult to absorb, indicating an incomplete or differential process of proletarianisation: they were seen as a lumpenproletariat in this still poorly industrialised part of the world (Bozzoli 1987, p. 23; Van Onselen 1979). The rise of gangs on the outskirts of Johannesburg in the late nineteenth and early twentieth centuries was viewed as the emergence of a black lumpenproletariat of landless farmers, nostalgic for rural peasant life and resistant to proletarianisation (Van Onselen 2001 [1982]. Jobless Koranic students taking up arms against the wealthy elites in Kano in 1980 were also seen as the urban poorest resisting proletarianisation (Lubeck 1985). We argue here that colonial metropolises offered a space for experimenting with measures of exclusion towards the youth population that were central during the compassionate period in Nigeria and the early apartheid period in South Africa. In line with the findings of Chapter 1, I suggest that the invention of new administrative categories (juvenile delinquents, isolated women, youth from the countryside, children in need of care, juvenile female street traders) were appropriated and supported by social groups that helped to implement them in their everyday actions. The multiplication of anti-delinquency committees in South African or Nigerian townships is one example among others of public action co-produced by bureaucrats and neighbourhood leaders. In South Africa more specifically, racialisation of youth policies, combined with massive incarceration, created a prison environment in which children who passed through the welfare system underwent violent criminal socialisation.

Children and youth protection policies were at the heart of a tension and an ambivalence that runs through the history of the penal system and protection of minors in the nineteenth and twentieth centuries in Europe as well as in Africa. In Western Europe, they accompanied the development of a welfare state that sought, at one and the same time, to 'rehabilitate' and help children 'in need of care' and to punish the most recalcitrant among them. These policies were also likely to deliver more young people into the hands of the criminal justice system than to give them support and social assistance (Muncie 1999, p. 257). The drift towards a discriminatory security policy and the criminalisation of minors that has taken place in the United States and some Western European countries in the past 30 years (Bailleau, Cartuyvels and de Fraene 2009; Muchielli 2008) has not ruled out more systematic management of children 'in need of care' by a state eager to engage its social workers in moral reform of the family by imparting the norms of individualisation, parental involvement, autonomy and gender equality (Serre 2009). Some view this process as a punitive turn in penal policy, as the sign of a process of criminalising the poor – with the fight against street delinquency as the counterpart and mirror image of the increasingly widespread insecurity of wage earners' working conditions – and the gradual imposition of a neoliberal political project in opposition to the welfare state model of the post–Second World War period (Wacquant 2004). This trajectory, understood as the withering away of the social state and, in its place, the expansion of the penal state – anyway a highly debatable interpretation (Carrier 2010) – does not square with the history of state formation in Africa (Hilgers 2012). To explain the fundamental variations in welfare policies and forms of punishment, François Bonnet (2019) suggests instead to use the theory of less eligibility: 'every society has to make welfare less attractive than minimum wage work and arrange punishment so that crime is less attractive than welfare. The living standards of the lowest class of workers in a society determine the upper limit for the generosity of welfare and the humanity of punishment in that society' (Bonnet 2019, pp. 5–6). In other words, young people dealt with by social welfare services could not be given more generous treatments than casual young and poor workers. Welfare colonial services because they were especially poor were also fundamentally coercive towards those they initially planned to protect. This chapter explores this in detail.

The urban socialisation of children and youth became a source of renewed concern in the 1930s and 1940s, leading to the production of further knowledge and new policies on children. The new penal standards were devised under the competing influence of experts and the immediate preoccupations of governments and administrations, and were promoted by divergent interest groups. In both Nigeria and South Africa, protection policies were designed and implemented mainly by one department (social welfare), which claimed to be reformative. Juvenile delinquents and children in need of care were 'naturalised' by a newly defined legal age of children and youth,[1] but above all by a bureaucracy that made the two groups a clearly identifiable social problem requiring specific policies. By re-examining the particular moment when these policies first arose and were later reversed, we can question what this reversal of public action policy actually meant in the colonial context, and grasp in detail its makeshift solutions and the haphazard circumstances under which they came into being. A close analysis allows us to reveal the original ambivalence of this action followed by its almost immediate reversal. We see the embryonic development of a state that purports to be protective into a late colonial state in Nigeria and a racialised welfare state in South Africa which can be both described as coercive because they exercised considerable violence on the extremely vulnerable populations they initially intended to protect. We aim to demonstrate that the reform-minded, compassionate and modernising project of the late colonial state in Nigeria and the South African government actually provided the conditions for a more systematic intrusion upon groups whose individual behaviour they planned to reform. Before looking at the criminalisation of young urbans by social services during the 1950s and 1960s, there is a need to explore why and how urban poverty and delinquency issues became political issues in Nigeria and South Africa.

Rise of Urban Poverty and Delinquency Issues

In South Africa, the Children's Act of 1937 concerning childhood and youth came out of the intersecting influences of penal reform and a branch of applied psychology called psychometrics, which aimed to measure the 'mental differences' or intelligence of individuals. In particular, the law encompassed 'children in need of care' and 'juvenile delinquency' in a single analysis, thereby assigning a common origin to the problems of these two groups: the difficulties faced by families in adapting to the country's new industrial and urban environment. Indeed, the law's definition of 'children in need of care' testifies to the absence of any clear-cut boundary between a 'delinquent child' and a 'child in need of care' precisely because indigence and delinquency were perceived as one and the same problem. According to the law, 'a child in need of care' was defined as 'an indigent child, a child with no parental control, a child that kept immoral company, a child that had committed an offence, a child beggar or a child engaging in street vending'.[2] As indigence and lack of family stability led to delinquency, the overall philosophy of the law was therefore to improve the child's family living conditions. It provided for the establishment of children's' courts (up to 19 years of age) and juvenile courts (between 19 and 21 years of age); the creation of a body of child protection and probation officers responsible for finding an alternative to reformatories and prison (approved schools, adoption, placement under the guardianship of probation officers) (Fourchard 2011c). The law contained two main sections designed to prevent children from becoming 'socially maladjusted' or delinquent. The first step was a set of measures enabling the government to intervene when children were considered 'neglected' and to 'rehabilitate' those considered delinquents. 'Rehabilitation' meant that taking charge of children 'in need of care' and delinquents became an educational rather than a criminal matter. It involved removing these children from their environment and putting them under the supervision of social services, particularly a probation officer working in tandem with child protection agencies. The spirit of the law was summed up by its chief advocate, the then minister of education, Jan Hofmeyr, in an address to the Parliament in March 1937:

'Fundamentally implicit in this law is an essential unity between the problem of juvenile indigence and the problem of juvenile delinquency.... . The two things are aspects of one and the same problem ... deterrent measures, punitive measures are not enough ... we need a constructive programme, not merely a programme of penalties, not merely a programme of threats, but a programme which will consist of educational measures and of social rehabilitation factors, and which will consist of psychological factors and mental hygiene factors.'[3]

This speech reveals the influence of two new spheres of expertise in the social sciences: the first, a group promoting penal reform and social action in favour of Africans, and the second, a group of psychometricians concerned about 'problematic' white children. Indeed, the law on childhood and all its original ambiguity must in fact be understood as a reaction to the emergence of a twofold issue at the core of political debates in the 1930s – the problem of 'poor whites' and the problem of the proletarianisation of black populations. During this period, a number of ministerial and inter-ministerial reports as well as sociological and anthropological analyses by academics and independent associative groups mention the widespread development of African urban poverty, marked by what was then called growing family 'destabilisation' (an increase in the number of single women, 'without husbands' or 'not really married'), and the inability of municipalities to halt the influx of rural populations into the areas under their authority (Posel 2005). This observation generated political concerns, propelled the issue of the deteriorating living conditions of black city dwellers into the spotlight, and paved the way to a more systematic analysis of the aetiology of deviance among African youths and children (Glaser 2000, p. 21; Van der Spuy, Schärf and Lever 2000).

This social question was preceded by another, even more nagging issue in white parliamentary circles: the problem of 'poor whites'. The massive rise in the number of poor whites in the 1920s prompted the government to launch a wide-scale investigation between 1929 and 1932: the Carnegie Commission of Inquiry on the poor white problem in South Africa. The inquiry did not lead to the creation of a welfare state in South Africa as some researchers have claimed, because the first social reforms had taken place before the commission was set up (Seekings 2008). However, by identifying 300,000 poor whites (i.e. 17% of the white population), the commission made the problem of poor whites a national question, which then became a partisan issue among opposing parliamentary political forces and a major impetus for the mobilisation of nationalist Afrikaners before and during the Second World War (Giliomee 2003, p. 347). In this context, the position of the children of poor white families became the subject of a self-proclaimed modern scientific study, notably by a new body of psychometricians.

Between Psychometric Expertise and Penal Reform in South Africa

The inter-ministerial committee that prepared the revision of the 1937 law on children and the subsequent measures in favour of children attests to the widening influence of South African psychological expertise among policy-makers. During the interwar period, American psychometric methods were popularised in South Africa by a few key figures through the frequent use of standardised IQ tests on children, whether they attended school or not, and from our standpoint, children 'in need of care' that had been through the social services and children's court system. The tests, developed between the late nineteenth and early twentieth centuries in France, England and the United States, quickly became a way to divide a given population (pupils, students, soldiers, migrants) into groups classified as 'gifted', 'average', 'backward' and 'moronic'. Bolstered by a technocratic rationale, this approach made it possible to detect exceptionally intelligent pupils and diagnose deviance or various social problems such as academic failure, unemployment or delinquency. It immediately aroused heated controversies that were to last for decades in the United States and Europe regarding its methodological validity, scientific integrity and underlying political motives (Montagus 1975).

IQ tests were introduced in South Africa in the mid-1920s by a handful of South African academics trained in the United States (Malherbe 1977, pp. 316–317) but it was not until the 1930s that a small group of psychometricians acquired nationwide influence and had a significant impact on the treatment of children 'in need of care'. At the time, this group included only white children.[4] Three personalities in particular – Ernest Gideon Malherbe (1895–1980), Raymond William Wilcocks (1892–1967) and Louis van Schalkwijk (1876–1961) – put their knowledge to work carrying out major governmental surveys on poverty, on indigent and delinquent children. In preparing the Carnegie report, Malherbe and Wilcocks (university lecturers in education and psychology, respectively), were the first to conduct intelligence tests on several thousand white children and assign them intelligence quotients: so-called 'average' children had an IQ above 89; subnormal 'stupid or mediocre' children had IQs of between 89 and 69; 'children with psychopathic tendencies' or 'morons' had an IQ below 69 (Malherbe 1932, pp. 185–186; Wilcocks 1932, p. 147). Both academics advocated creating specialised classes or schools reserved for 'subnormal' and 'pathological' children that would teach them to become skilled workers suitable for future industrial employment. The report was decisive in shaping how the issue of childhood indigence and delinquency was understood and in familiarising pupils with intelligence tests and expanding their use on schoolchildren by the National Bureau for Educational and Social Research set up by Malherbe in 1929 (Fleisch 1995).

A few years later, concerned by the increasingly precarious situation of African urban families in townships, Jan Hofmyer, the new liberal minister of Education and Health, created an inter-ministerial committee on poor and delinquent children headed by Louis van Schalkwijk, an inspector for the Ministry of Education.[5] The committee's report provided the basis for discussion regarding the new law on children. Van Schalkwijk, who had a doctorate in applied psychology from the University of Amsterdam, had become aware of psychometrics through Malherbe's National Bureau and his own research for the South African Group of Intelligence. His report stressed the need to match treatment of 'children in need of care' to their intelligence and their classification as 'normal', 'subnormal' or 'pathological'. The testing of white children in reformatories, begun in the early 1930s, was henceforth used more systematically by the Ministry of Education (Chisholm no date, p. 17). After the law was adopted in 1937, white children under the supervision of social services were classified according to their IQ test results, and the Ministry of Education consequently classified its reformatories and approved schools under three headings: 'normal', 'subnormal' and 'psychopathic'.[6] This classification system testifies to a newfound interest in individualised treatment of indigent children, but the decision to base it on IQ tests led above all to a process of creating 'unitised subjects', to borrow Megan Vaugan's expression, which naturalised them as distinct categories. Due to a more systematic testing of children 'in need of care' and their low scores compared with the white national average, South African experts at the Ministry of Education concluded that intellectual deficiency was an essential factor in delinquency and indigence within the country. This correlation dovetailed with the proposals put forth by the English psychometrician Cyril Burt, whose view of intelligence as a hereditary rather than acquired trait had an early and lasting influence on South Africa.[7] Although psychometrics enjoyed increasing recognition within the Ministry of National Education, IQ tests were seldom used outside white institutions.[8] Furthermore, they were by no means the only solution under consideration within the Ministry of Education, which was also attracted to the penal reform movement.

In the early 1930s, the theories of urban–rural disconnect and the social dislocation caused by the urban environment formed the most widespread dogma used to account for the rise of African juvenile delinquency in South Africa, an explanation that tended to embellish rural life and consider 'the city' its undesirable opposite (Glaser 2000, p. 45). An alternative interpretation emerged from a reform movement, which included the South African Institute of Race Relations (SAIRR) set up in 1929, the Johannesburg Non-European Affairs Department (JNEAD), the new Union Ministry of Social Affairs created in 1935, the Committee and later the League for Penal Reform founded by the SAIRR in 1940, as well as a number of directors of approved schools such as Alan Paton, director of the Diepkloof reformatory (1934–1948), and William Marsh, a judge at the Johannesburg juvenile court (1938–1945). Ellen Hellman, an anthropologist at the University of Witwatersrand, was probably the most well-informed commentator on this issue (Hellman 1940, 1948). She wrote profusely about urban children and youth and was one of the first anthropologists to deconstruct commonly accepted notions of detribalisation of Africans. These notions, which were widely employed in administrative circles, viewed all African migrants as detribalised persons, i.e. cut off from their rural roots and the authority of the so-called 'traditional chiefs'.

This milieu was closely connected to the Minister of Education at the time, who was responsible for the passage of the Children's Act of 1937. Hofmyer shared the views of Alan Paton and Ellen Hellman, both of whom campaigned in favour of specialised education for African children in need of care.[9] By advocating alternatives to imprisonment and analysing delinquency and the neglect of youth and children as a single problem, the new law broadly reflected the influence of the country's liberal, welfarist current as well as the impact of the 1933 British law on children that served as a partial model for the 1937 law in South Africa (Fourchard 2011c, p. 522). In 1938, this group of scholars and educationists took the initiative to hold a conference on African urban juvenile delinquency in Johannesburg, which for the first time presented a comprehensive, detailed analysis of the primary causes of the delinquency and neglect of youth and children (Glaser 2000, p. 24; SAIRR 1938). The group's consensus on poverty, lack of housing and education and the prevalence of monoparental families as the core of this social issue invalidated the theory of urbanisation as the cause of social disorganisation. It argued, on the contrary, that urban citizenship should be granted to the majority of Africans migrants, who had hitherto been prohibited from residing in the city, except for the short period of their labour contracts (Chapter 1). It insisted on the need for urban socialisation, i.e. to provide them with decent living conditions and improve their access to education and basic services, rather than consider them *a priori* as individuals in the process of becoming detribalised (Posel 2005, p. 75). Finally, it called for more active state involvement, increased spending on social needs (housing, education, leisure activities), and higher wages to eradicate proletarianisation, delinquency and the threat to children posed by such an environment. In the wake of the Johannesburg conference, this liberal current took action somewhat like the Howard League, the principal reform lobby in the British government (Ryan 1978), setting up a monitoring committee to pressure the government into demanding broader implementation of the 1937 law and militate for effective adoption of the principal recommendations of the 1938 conference. They made sure their efforts received national and international attention by publicising them in various academic journals and activist pamphlets (Junod 1948; March 1946; Paton 1948).

The Empire's First Social Services in Lagos

South Africa was not the only country to undergo the influence of the penal reform movement during the 1930s, but of all the African countries, it was undoubtedly where the greatest strides were made, thanks to the efforts of activist associations, committed academics and liberal civil servants. Elsewhere, the problem of indigent and delinquent youth was a marginal issue, particularly in the British and French colonies; at best, it was beginning to draw attention in public debate in countries such as Kenya and Tanzania (Burton 2001, p. 200; Campbell 2006, pp. 142–143). The penal reform movement progressed slowly in the Colonial Office. In 1930, a draft law recommended that the colonies introduce specific provisions in their penal system (juvenile courts, remand homes and the creation of posts for probation officers). These requests were ignored in the empire's African colonies.

In Nigeria, such an initiative seemed pointless, given that, according to the governor general, only a negligible number of juveniles went before courts. The law provided for their segregation in prisons, whereas the construction of reformatories and probation officers' salaries were judged too costly during a period of budget restrictions. Social problems, poverty and delinquency were either ignored or dealt with by Christian missionaries, the family or supposedly traditional African solidarity. Only one approved school was opened in 1937 in Enugu in the East of Nigeria. Donald Faulkner, a civil servant in the British penitentiary department and zealous promoter of penal reform, was named its director (1937–1941). In 1940, no one in the colonial administration saw the offences committed by minors or the precarious living conditions of poor children as a problem. The authorities were deaf to the continual complaints of Lagos residents, who had been denouncing since the 1920s the wrongdoing of *jaguda boys* and *boma boys*, groups of young boys specialised in techniques of pickpocketing, purse snatching and assaults by local gangs (Fourchard 2006a, pp. 123–126; Heap 2011).

Unlike in South Africa, where the passage of the law of 1937 was preceded by long, preliminary studies, the issue of juvenile indigence and delinquency suddenly burst upon Nigeria by chance at the start of the Second World War, giving reformers at the Colonial Office a pretext to step up the penal reform movement within the empire. Lagos was transformed into a base for the British Navy, prompting *boma boys* to become instant 'tourist guides' for the thousands of soldiers passing through, whom they robbed as they came out of nightclubs and brothels. These actions, seen by the governor general as 'a betrayal of the fatherland in wartime', led him to decree a ban on 'unlicensed' guides in Lagos or at least to ensure that children and youths most exposed to urban risks and danger were taken in charge. Faulkner's report, based on the first studies on this milieu, was decisive and was to play a crucial role in structuring social services later on. The social services reports from Calabar and Onitsha (in the East Region), Ibadan (in the West Region) and the Lagos federal services after independence[10] confirmed Faulkner's observations, but it is hard to tell whether their conclusions were the product of bureaucratic routine or lack of critical distance of the second generation of British civil servants and the first generation of Nigerian civil servants that had trained them since the 1940s.

In 1943, Faulkner encouraged the governor general to pass the Children and Young Person Ordinance (CYPO), a Nigerian adaptation of the British law of 1933, which sought to avoid prison for minors and instead put them in separate remand homes from those of adults, dissociate delinquent children from those in need of care and, depending on the case, send them home to their parents or place them in specialised institutions. The primary goal of this law, like the law of 1937 in South Africa, was to avoid the crime-inducing effects of prison on adolescents and children. During a visit to Nigeria the same year, Alexander Paterson discovered that 80% of adolescents and children who went before courts were either flogged before being released or imprisoned along with adults.[11] He supported Faulkner's desire to introduce social services capable of handling these groups separately, a goal he achieved in 1944 by setting up the social welfare department, opening a juvenile court in 1946 and creating a special police force to escort, investigate and return youths to the countryside. From that year onwards, the statistics on juvenile delinquency were compiled separately in the annual reports of the police and the welfare service suddenly constituting the legal and administrative reality of a long-standing phenomenon. Thus juvenile delinquency became a distinct social problem dealt with specifically by the colony's social services. The few reformers in Nigeria and the Colonial Office were in the avant-garde of this action. Whereas penal reform had stalled during the 1930s, they succeeded this time, thanks to the war and its momentary obsessions – urban disorder and delinquency (Lewis 2000, p. 21) – in imposing their agenda based on the belief that state intervention could resolve the empire's social problems.

Race, Gender and Welfare

In both countries, the new system created the conditions for a very systematic state intervention in the lives of poor families in urban areas. Unsurprisingly, this intervention was far greater in South Africa than in Nigeria, particularly with regard to youth between the ages of 17 and 21 (Tables 2.1 and 2.2). In South Africa, 21,500 children and youth went before the courts every year during the mid-1930s; 20 years after the law was implemented, the number had risen to 177,923 (i.e. multiplied by eight), whereas the population had only doubled during the same period (from 8 to 16 million).[12]

TABLE 2.1 Population under age 17 who went before a children's court

	South Africa	Nigeria
1945	16 550	220
1962	35 483	5 000*

Source: Based on South African Republic (1963).
This figure of 5000 is uncertain: it is an extrapolation at the national level of a few figures obtained in a few children's courts of the country (Lagos, Kaduna, Enugu, Calabar, Ibadan).

TABLE 2.2 Population under age 21 who went before children's or juvenile courts

	South Africa
1933	21 536
1945	77 604
1957	177 923

Source: Based on South African Republic (1963).

Aside from more intensive state intrusion in everyday family life, the welfare services in the two countries developed different orientations. In post-war Nigeria, the protection of minor girls became an issue that exacerbated the divisions between the supporters and opponents of social service policies. In South Africa, racial preference policies, stemming from the allocation of different resources for children and youth according to their racial classification, radically transformed the spirit of the law of 1937.

From Preference to Racial Differentiation in South Africa

The period of 1937–1948 represented a moment of political possibilities in which the social science expertise in the service of the government was marked by contradictory influences and attempts at reform (Dubow and Jeeves 2005). This period must be distinguished from the subsequent triumph of apartheid (1948–1976), characterised by the quashing of alternative liberal proposals and the simultaneous assertion of an ideology to serve a policy of so-called 'separate development of the races' (apartheid).

But from 1937 to 1948, child protection policies were ambivalent, reflecting the division within the governing coalition between the United

Party, some of whose members wanted to limit racial discrimination, and the Afrikaner nationalists who wanted to strengthen it. Overall, the period was favourable to the reformist child policy pursued within the government by Jan Hofmyer (Minister of Education and Public Health from 1933 to 1938, Minister of Education and Finance from 1939 to 1948) and in the field by a number of reformatory directors and children's court judges who became the pioneers of rehabilitation. The first approved school for coloured children, set up in the outskirts of Cape Town in 1948, was modelled on those for white children, a testimony to the desire to extend rehabilitation policies to other racial groups through individualised treatment on a case-by-case basis (Badroodien 2001, p. 4). Alan Paton, the director of the country's sole reformatory for African children (the Diepkloof reformatory), transformed its penal and military discipline into a 'school' that included diversification, differentiation and individualised treatment (Chisholm 1991). He set up a hostel for rehabilitated young delinquents working as day labourers in Johannesburg; he introduced a system of gradual release and had more faith in ritualised obedience to authority than in the systematic use of corporal punishment (Paton 1948, 1986). William Marsh, the first probation officer for African children in Johannesburg, declared that the idea of punishment had been almost totally abandoned by the juvenile courts in Johannesburg, but he acknowledged that a reasonable amount of flogging was a permissible substitute when no facilities were available to receive the children (March 1946, pp. 24–26).

Nevertheless, the actions of the rehabilitation advocates cannot hide altogether the ambivalent effects of the law of 1937. A report in 1947 noted that the implementation of legislation regarding children was still largely experimental and that it had mainly benefitted the white minority:[13] 20,000 white children had been accommodated or placed in a facility, compared with fewer than 8,000 African children. White children almost always found a place in the institutions provided for them (foster families, charitable associations, approved schools). For Africans, the effects of state intervention were completely different. Reformatories were scarce and overcrowded, and reducing detention periods became a solution to keep the headcount in check, thereby compromising any likelihood of medium-term rehabilitation. Institutional placement was in fact the exception: in 1945, it concerned only 2.5% of the African minors who went before the courts.[14] Outcomes varied depending on the judge's clemency, the offence committed, and the child's situation, ranging from a mere reprimand to a prison sentence for rape or homicide. Stealing was by far the most common offence, and before being sent home to their families, offenders were usually punished with a few lashes of the whip, a practice that social workers considered preferable to a prison stay.

The unequal resources invested by the South African government thus result-ed de facto in a policy of racial preference: rehabilitation and individualised treatment for white children; collective penalties and the use of corporal pun-ishment for African children.

This process was reinforced by the arrival of the National Party at the head of the South African government in 1948. Institutions offering medical and psychological care and approved schools were reserved for white children by the centre of psychometric specialists, whereas penal solutions were more systematically implemented for African children by judicial and prison insti-tutions and 'racial' ministries (the Bantu Affairs Department, the Coloured Affairs Department, the Indian Affairs Department). The racial preference policies based on unequal financial investment, which reformers condemned and the government had trouble justifying, were replaced by a policy of racial differentiation, which separated white children suitable for rehabilitation from African children henceforth classified as dangerous because of their proclivity for idleness (Seekings and Natrass 2006, p. 169). All efforts undertaken by the United Party government were pursued on behalf of white children: 16 approved schools for children in need of care and two reformatories for delin-quents completed the institutional structure designed to cover all the school-ing needs of the white population at the end of the 1960s. The government had clearly chosen to invest in the education of poor whites to ensure that the non-competitive positions of white parents in the labour market would not be passed on to their children (Seekings and Natrass 2006, pp. 162–164).[15]

In contrast to the pre-apartheid period, this policy was legitimised by the now dominant expertise of Afrikaner criminologists who justified dif-ferentiated racial treatment on the grounds of supposedly different 'crime-inducing' behaviours rooted in race. These experts arrived at their analysis of delinquency by combining two opposing interpretations in the US: the Chi-cago School theory of social disorganisation and psychometrics. In essence, the analyses of urban sociology produced by the Chicago School during the interwar period held that delinquency stems from the social disorganisation of cities, as well as the decline of social norms (e.g. the decreasing influence of behavioural rules on members of a group) and the lack of stable social relationships in the poor neighbourhoods in American city centres, partic-ularly in Chicago, due to the rapid mobility of migrations (Burgess 1926; Shaw 1929; Shaw and McKay 1942). Apartheid criminologists adopted the theory, perverting it with a localised version of social disorganisation (Van der Spuy, Scharf and Lever 2000). They suggested that each racial community was affected in a different way by exposure to a naturally crime-inducing urban environment: young Africans challenged laws and authority in the city because they were detribalised; the young coloureds because they straddled the line between whites and blacks, with no particular original culture, which led them to commit many more offences than the average in the other groups; finally, the whites because they were too exposed to material values and the cult of individualism specific to Western city life (Venter and Retief 1960). For example, a study conducted on behalf of the National Bureau for Educational and Social Research by the criminologist Jacobus Venter of the University of Pretoria was able to marshal judicial statistics to establish correlations between racial lifestyles and predominant types of criminal behaviour: young whites violated road safety rules, young coloureds committed assaults, and young Africans engaged in thefts of less than 50 pounds (Venter 1959). This racialised interpretation became pervasive just as the penal reform movement, which had been so dynamic during the 1940s, was running out of steam.[16]

In the 1930s, control over juvenile institutions had been transferred from the Ministry of Justice to the Ministries of Education and Social Affairs, and paved the way for a new generation of school directors and judges favourable to child rehabilitation. Under apartheid, only white children continued to be supervised by the two ministries, which separated approved schools for intelligent children from centres for mentally retarded or subnormal children. Children's intelligence was regularly tested to determine whether or not they were improving in these facilities (Kaldenberg 1951). Psychometrics and its individualised treatment methods ultimately prevailed at the Ministry of Education and monopolised the field of supervised schooling.[17] On the other hand, at the end of the 1950s, African, coloured and Indian children were placed under the authority of the new 'racial' ministries. This institutional change signalled the end of the reformist spirit, eliminated the liberal option, and put into practice the core idea of the nationalist criminologists who believed that each racial group was distinguished by specific behaviours and consequently required differentiated penal or educational responses. However as racial affiliation became key in South Africa, gender and the forms of juvenile prostitution in particular became a major topic of reflection and measures initiated by the social services during and after the Second World War in Nigeria.

Juvenile Prostitution and the Construction of a Moral Space in Nigeria

The fight against juvenile prostitution contributed to the historical construction of the notion of intolerable conduct, shifting the boundaries of moral space in contemporary Western Europe beginning in the nineteenth century (Fassin and Bourdelais 2005). In France and England, juvenile maltreatment and child labour gradually became a category of social misery constructed by institutions, experts, charitable associations and a certain philanthropic press (Bourdelais 2005). Juvenile prostitution underwent a similar process owing to the violence against bodily integrity and sexual abuse the children suffered. It was obvious in Britain where, as a result of a series of newspaper articles and associations and churches engaged in anti-child abuse activities, a set of laws was passed between the late nineteenth century and the 1930s to protect minors from sexual relations that were non-consensual, transactional or facilitated by an 'amoral' environment (Brown 2004). It was also apparent at the international level: since the 1870s, a worldwide campaign against the traffic in women culminated in the international convention for the eradication of 'white slavery' in 1910, a struggle relayed by the League of Nations starting in 1921 and later by the United Nations (Limoncelli 2010).

In Nigeria, juvenile prostitution was criminalised during the Second World War, less as a result of pressure from 'civil society' than from an article in a colonial scandal sheet. In the colonial context, the issues that prompted the promulgation of this penal norm differed from those stated by its protagonists and those who controlled its drafting in the UK. Similarly, the system introduced after the Second World War referred to the particular conditions under which the norm was constructed: although it attested to an apparent determination to protect minor girls, it also indicated first to the British and then the Nigerian authorities its willingness to accommodate existing forms of procurement in Nigeria.

The British authorities had known about juvenile prostitution and the sexual exploitation of young girls in Nigeria since the 1920s; they nevertheless denied its existence before the League of Nations and asserted, on the contrary, that the only surviving forms of pre-colonial slavery in the colony were found in domestic service (Aderinto 2012). In the 1930s, several administrative reports noted 'trafficking' of girls and women between the Obubra district in Calabar province and the main ports in Nigeria (Calabar, Lagos), the Gold Coast (Accra, Sekondi and Takoradi), and Equatorial Guinea (Fernando Po) (Naanen 1991). Prince Eikineh, leader of the Ghanaian branch of the Nigerian Youth Movement (NYM), sounded the alarm in 1939 and delivered concrete proof to the local authorities of a traffic in young girls between Nigeria and the Gold Coast, in which girls were forced into prostitution by older Nigerian women.[18] The action taken by the NYM at the time was relatively independent of any resolve to relieve the social misery of the girls involved. It reflected above all an eagerness to avoid tarnishing the moral reputation of the Nigerian community in Ghana. Tribal unions and home town associations worked together with the authorities to ensure that any Nigerians whose activities were perceived as a threat to the success of their community would be sent to prison or expelled (Aderinto 2012). They had good reason to be concerned. Prostitution was so widespread in certain groups such as the Akunaakuna in Calabar province that the terms *akunaakuna* or *akuna* became synonyms for prostitution and prostitutes for many populations in the Eastern Region of Nigeria as early as the 1940s (Ekpo-Otu 2013).

The colonial authorities tolerated the existence of this traffic until 1941, when two articles appeared in *West Africa*, a colonial magazine widely disseminated in Great Britain and Africa, citing some of the information presented by Prince Ekineh and denouncing the inability of the British authorities to stop the traffic between Nigeria and the Gold Coast (Aderinto 2012, pp. 13–15). Soon after the flurry of media attention and the arrival of numerous petitions at the Colonial Office, an amendment to the penal code of the Gold Coast and Nigeria was passed, making the prostitution of minors under age 13 a criminal offence and prohibiting hotels for prostitutes, brothels and guardians who assumed temporary parental authority.[19] The Colonial Office was thus responding to sudden international pressure by demonstrating its active participation in the fight against human traffic. It was also responding to the will of British penal administration reformers like Paterson, who recommended the appointment of social workers to combat Gold Coast trafficking.[20] The 1946 appointment of Alison Izzett as the Lagos social welfare department's first social worker was in line with this goal. This 'moral panic' over imperiled girlhood sexuality well reflects this new moral anxiety by the British government about the best means of protecting children and especially juvenile girls (Aderinto 2014, pp. 93–102).

In Nigeria itself, the decisions made by the administration and social welfare services managers had however little effect on procuring or combatting juvenile prostitution. For senior welfare officers in Lagos, Donald Faulkner and Alison Izett, three factors accounted for the prostitution of minors: sham or what was referred to as *marriage by proxy*, street trading which was conducive to accepting sexual advances, and the housing of girls by guardians, who took advantage of the situation to exploit them. To engage this fight, social services could count on the various associations in Lagos that were committed to eradicating prostitution for other reasons. Associations of immigrants from the Calabar and Owerri provinces were calling for energetic measures against prostitutes, and by extension, against all the single women who were coming to Lagos and 'ruining the good name of their ethnic group'.[21] Since the 1920s, the Lagos Women's League, an association of educated Nigerian women that promoted education for girls, had been calling for stricter regulation of street trading, due to its attendant moral dangers for girl (Georges 2011). With the development of the welfare service, a subsection of the 1943 Children and Young Person's Ordinance eventually came to prohibit all children under fourteen from selling petty goods in the street (Georges 2014, pp. 108–109). Girls between fourteen and sixteen were banned from hawking in the central business district and the vicinity of bars, brothels and military barracks: the ordinance sought to impede sexual contact between young girls and men (Georges 2014, p. 109). The welfare office also barred young women from travelling to the capital unaccompanied. The two measures were implemented with the help of the police and associations in the prostitutes' native regions, in spite of the objections and repeated demonstrations of Lagos women traders.

The *West African Pilot* newspaper and the National Council of Nigeria and the Cameroons (NCNC), which defended the interests of the street traders in the 1940s, denounced the violation of freedom of movement and employment, indiscriminate arrests, the invasion of privacy, the heavy fines imposed on poor families and the fact that the female street vendors were henceforth required to reside in a hotel along with real prostitutes. Despite the protests, first the colonial social services and later the Nigerian federal government upheld the measures until the Biafran War (see below). Of course, the criminalisation of the street economy was prompted by a desire to protect minor girls, but it failed to curtail juvenile prostitution. The reason was simple: of the three factors identified by Izzett and Faulkner, marriages by proxy were the most effective method of procuring in Nigeria. In 1943, Faulkner discovered that juvenile prostitutes were controlled by men or women pretending to be their husbands or guardians, but who were in fact pimps that paid a 'dowry' to the poor families of the girls in the provinces (mainly Calabar and Owerri) in exchange for a promise of marriage, employment and/or schooling in Lagos.[22] A new series of investigations carried out in 1946 brought further evidence: several Lagos nightclubs managed by older Nigerian women regularly offered their customers the services of girls between the ages of 12 and 15. These forms of procuring were partly an outgrowth of human pawnship, which became a widespread practice in the late nineteenth century due to the scarcity of manpower to meet the growing needs of the colonial economy (Falola 2003). Here we would argue that marriages by proxy were originally a form of bonded labour involving underaged girls used by their parents as collateral or 'pawns' to secure their debts (Fourchard 2013).[23] When pawnship was prohibited in 1927 for minors under 16 years of age, it seems to have been transformed into a sham marriage: the young girl was pledged illegally by her parents to a third party in exchange for a large sum of money. In this 'marriage', the contracted debt was replaced by a 'dowry' paid by the third party (usually a former prostitute) who promised to marry off the girl, most often in Lagos, Calabar or Accra. In turn, the girl was required to work for the third party until the debt contracted by her parents was fully repaid.

Fighting prostitution in Lagos and Nigeria therefore implied regulating the institution of marriage. After the war, under the influence of the social services, the governor general in Lagos suggested prohibiting young girls from becoming engaged and marrying before the age of 18. In the face of opposition from the customary authorities in the provinces, who were loath to see to their control over the 'matrimonial market' restricted, legislation on the matter was dropped.[24] The expansion of social services in Lagos and the control of street trade showed the world that the British authorities, with the support of philanthropic associations, were committed to combatting the exploitation of minor girls. However, the measures adopted were either ineffectual (the ban on girls under age 16 coming to Lagos unaccompanied was soon abandoned because it was impossible to implement) or misdirected: the prohibition against street trade in no way limited existing forms of prostitution. Taking measures against the various forms of proxy marriages meant alienating customary authorities, which colonial officials refused to do. They therefore turned a blind eye to these disguised bondage practices, which were condemned by the international community and were to enjoy a bright future in the case of Nigerian female trafficking networks.[25] In short, from the colonial point of view, juvenile prostitution was only *relatively* intolerable. Instead the limitation placed on the freedom of children and young girls to move about and work became part of a systematic coercive policy against urban youth after World War II.

A Coercive Incomplete Welfare State

In Chapter 1, I have already mentioned that the stabilisation policies of the first decade of apartheid were accompanied by a measure limiting access to the city by migrants. The stabilisation policies in the British colonial empire were also accompanied by similar restrictions on the so-called undesirable populations, but on a larger scale. In the new post-1945 order obsessed by the need to stabilise a class of urban workers, officials showed greater determination to separate the hardworking from the lazy, workers committed to their trades from passing migrants and supposedly orderly groups from dangerous populations (Cooper 1995, p. 262): the unemployed, delinquents, single women and unemployed youth came under increasing pressure. More than ever, young people in particular had to work either in school or elsewhere instead of loitering in the streets (Waller 2006, p. 41). The lack of schools and the limited workforce transformed these stabilisation and social assistance policies into coercion of the young: flogging, deportation to the country-side, massive imprisonment and forced labour became more systematic in South Africa under apartheid; fines, the use of the whip, restrictions on freedom of movement and employment and deportation to the countryside were extended from the Lagos colony to the three other regions in Nigeria by expanding social services.

Among these various punishments, flogging deserves special attention because it became a commonplace, accepted practice in both countries. It enjoyed institutional approval and was endorsed, as in many other countries on the continent, by social groups (association leaders, parents' associations, traditional authorities) seeking to correct behaviours viewed as deviant in relation to their own moral codes (Bayart 2008b). Whereas the central purpose of the reform was to avoid imprisoning minors, the absence of any alternative to prison due to the lack of budget and the phase of bureaucratic expansion, these states were undergoing favoured the institutionalisation of flogging. No doubt there was nothing original about this evolution: the increased reliance on corporal punishment of youth in colonial India was stepped up precisely as a result of the development of the modern state (Sen 2004). What was perhaps more unusual in the cases of Nigeria and South Africa was the fact that flogging became a distinctive way of dealing with the supposed unruliness of young people at the very moment when such forms of punishment were on the decline in the penal management of adults in both countries. Repressive policies with regard to youth and children were widely shared by parents, neighbourhood associations, elders' organisations and school teachers who became increasingly worried about the activities of gangs, delinquents, prostitutes and wandering youth (Kynoch 2003; Last 2000; Waller 2006).

This echoes the principle of 'less eligibility' introduced by François Bonnet (2019) to explain the fundamental variations in welfare policies and forms of punishment among Western welfare states. The living standards of the lowest class of workers in a society determine the upper limit for the generosity of welfare and the humanity of punishment in that society (Bonnet 2019, pp. 5–6). In both colonial Nigeria and apartheid South Africa the living standards of the lowest class of urban workers, and especially young people were very low. Simultaneously, punishment in the welfare system must be less attractive than crime. In other words, young people dealt with by the social welfare services could not be given more generous treatments than beggars or casual young workers. Limited welfare services in colonial states because they were poor were also fundamentally coercive towards those they initially planned to protect.

From Financial Indigence to Flogging in Urban Nigeria

In 1949, the Social Affairs Committee of the Colonial Office was convinced that, in view of its limited budget, it should concentrate its resources on a few target groups in city centres: children in need of care, juvenile delinquents and prostitutes.[26] This signalled a turning point in policymaking. Compassionate policies were expensive, but were only relevant to a tiny urban minority. Instead, the Colonial Office argued in favour of giving rural local development priority through the promotion of projects managed and financed by community development (Iliffe 1987, pp. 202–207). Tanzania is a good example of this policy conversion: participation of the local population, promoted by the social services, became an integral part of development projects, and continued to shape them during the first decade of independence (Eckert 2004; Jennings and Mercer 2011). Nigeria was an exception to this imperial evolution. Not only did the Lagos social services retain their initial focus on urban youth, but the policies developed by the new services set up in the three regions pertained exclusively to urban youths until the Civil War (1967–1970). The examples of federal social services in Lagos show that rehabilitation projects directed at young offenders were limited by tight budgetary constraints, whereas political leaders constructed young people as a threat.

In 1950, the Lagos services had 8 European civil servants and 33 African assistants, whereas the other 3 regions of Nigeria had to make do with 3 European civil servants.[27] By 1960, the federal Ministry of Social Affairs in Lagos was handling 2,000 cases of youth and children 'in need of care' and youth per year, with some 30 volunteer workers coordinating the various federations of clubs and associations of young people, women and assistance for the disabled and the poor.[28] The reformist attitude was still percolating through the institution. Adolescents who had committed repeated offences could be accepted at one of the approved schools in the capital, which had a combined capacity of 250 children in 1960 (Fourchard 2006a, p. 135), i.e. half the slots for the entire country, which was hardly negligible in the context of scarce available seats in classrooms. In the early 1960s, Nigerian civil servants noted that parents came to report fictional offenses committed by their children so they could be admitted to one of the approved schools and receive housing and free professional training.[29]

Most of the energy of the social services were taken up by coercion, however. Their managers ignored the warnings of Colonial Office inspectors who reminded them that controlling street vending kept them away from achieving their goal of rehabilitating adolescents.[30] The federal government and the NCNC, which had backed street traders against the social services in the 1940s, now used social welfare and the police to weaken the local opposition and the traders of Lagos. The evolution of the juvenile court, from its creation to the time of the Civil War, was symptomatic of this change. In the 1950s, the court dealt with about 1,000 cases a year; during the first half of the 1960s, the number rose to 2,200–2,500 cases a year, i.e. 10 times more than the average of the other courts in the country. The increase in activity was mainly devoted to repressing hawking and street vending by minor girls. Although the number of offenses (theft, robbery) remained unchanged in the 1950s and 1960s (around 300 per year), the rate of hawking offences multiplied six-fold between the 1940s and the 1960s (from 200 cases a year between 1946 and 1951 to 1,250 cases a year in 1963 and 1964 (Fourchard 2011c, p. 529; Georges 2011, p. 848).

This growth attests to a significant change in welfare service. As early as 1949, the head of social welfare service in Lagos, Donald Faulkner acknowledged that the fines paid by parents to free a daughter held by the social services contributed substantially to paying for court operations (Georges 2011, p. 848). In extreme cases (girls engaging in prostitution, children fleeing parental authority or children in need of care), welfare officers took charge either of finding the parents or offering an alternative solution (placement in another family or at an orphanage). This task became increasingly difficult in the 1960s. It was often impossible for civil servants to find parents in the densely populated outskirts of Lagos where there were no street addresses.[31] The deportation of youth from Lagos to the countryside did not increase during the 1950s and 1960s because it had to be paid for by social services, which were reluctant to assume the expense – a very different approach from that of Tanzania where the authorities in Dar es Salaam were committed to a pro-active policy of deporting unemployed juveniles.[32] The efforts of the welfare services in Lagos therefore compounded their initial shortcomings: boys' and girls' homes became detention centres for minors who could not be released until their parents paid a mandatory fine. Proceeds from the fines were used to cover social service operating expenses. Under the First Republic, harassing minors became a method to force poor families to pay for the cost of implementing a policy supposedly designed to protect their children. Ultimately, this example reveals the government's ability to pass the cost of state apparatus directly on to the populations affected by its bureaucratic transformation.

Indeed, within the scope of social welfare regionalisation in Nigeria, funding quickly became the major obstacle to implementing compassionate policies. Abba Habib, the minister of social affairs in the northern region, regarded his ministry as the poorest of the government.[33] The idea of extending a welfare policy in the east had been suggested by the Ministry of Social Affairs, but in 1956, it was deemed too costly.[34] In 1960, most resources and efforts were absorbed by the creation of boys' and girls' homes and juvenile courts in the main cities of the east (Enugu, Aba and Onithsa). Of the young people who went before these juvenile courts between 1958 and 1963, 79% were sentenced to flogging, even though the eastern region had officially abolished the same penalty for adults in 1955 (Coldham 2000, p. 220; Milner 1972, p. 99). The ministry's financial indigence turned its initial reform-minded intent into a tool for disciplining youth. The difference in the treatment of adults and youth conveyed the profound conviction shared by members of the political class that flogging continued to be the most effective instrument to reform 'unruly' young people. In the northern region, the lack of facilities for receiving young people arrested by the social services, the police or native authorities meant that youth were either jailed in cells at police stations or barracks, or put in prisons along with adults.[35] As in the eastern and western regions, flogging was the most common punishment ordered by the juvenile court in Kano; there was a consensus regarding this practice among adults, where beatings were a common method of controlling youth and children at home and at school (Last 2000, p. 374).

Violent Socialisation of Urban Youth in South African Institutions

Ultimately, these coercive solutions proposed by Nigeria's new political and administrative elites were not very different from those recommended by the National Party for South Africa: preventing idle young people from entering the city was seen as a way to eliminate criminality and the supposed unruliness of urban youth – at least that was the government's official line for Africans (Van der Spuy, Scharf and Lever 2001, p. 173). As for white youth, the notion of 'children in need of care' retained its relevance, whereas for all other groups, it gave way to the notion of 'dangerous' or 'potentially dangerous' children because they were seen as idle (Seekings and Natrass 2006, p. 169). The apartheid government reached its disciplinarian and punitive turning point through a three-pronged policy shift involving the systematic use of forced labour, corporal punishment and massive incarceration. The South African government's ability to confine populations on a large scale and its concern about guaranteeing rural white voters a cheap workforce were the two major differences between its policies and those pursued in Nigeria. On the other hand, the whip – the low-cost instrument of youth justice in Nigeria par excellence – became, an essential vehicle of mediation between South Africa's institutions and its youth.

Within the framework of the Urban Labour Preference Policy and the need for labourers on white farms in the 1950s (Chapter 1), the Ministry of Native Affairs gave illegal Africans arrested in the city the possibility of accepting work in rural areas in exchange for the right to legal residence in the city at a later date (Posel 1991, p. 187). At the same time, an inter-ministerial survey revealed that 80% of young African men and women between the ages of 15 and 20 were living in the cities without employment (Seekings and Natrass 2006, p. 169). The department used this figure to send the 'idle' to 'Bantu youth camps' where they provided cheap manpower on the pretext of rehabilitation through work. The camps, whose history has yet to be written in detail, sprang up all across the country in the 1950s and 1960s under the name of 'youth service camps'.[36] Most boys arrested in urban areas were sent to these camps to work free of charge as day labourers on farms, where they supplemented the work performed by forced penal labour. This provision was later abandoned, when the mechanisation of white commercial agriculture in the 1970s made the need for manpower less urgent.

These repressive policies with regard to youth and children were viewed favourably in the townships among parents and elders' organisations that had been regularly calling for harsher measures since the 1930s to put an end to the danger of tsotsis (Kynoch 2003). Countless neighbourhood associations and anti-delinquency committees attested to rising insecurity and the development of a moral order that viewed youth as a danger to society, whose conduct had to be prevented and combatted (Chapters 3 and 4). The associations worked together with the police and social services to incarcerate or 'put to work' young people during the 1950s and 1960s. Like Nigeria, South Africa significantly increased the use of corporal punishment in institutions for minors. Reformers nevertheless sought to limit its use and thereby fight a tradition deeply embedded in the South African penal system. In 1946, the Lansdown Commission recommended that a delinquent minor should not be flogged if a more constructive form of punishment could be found (Midgley 1982, p. 395). The government ignored the recommendation and in 1952 an amendment to the penal code ordered the country's courts to impose mandatory corporal punishment on all those convicted of rape, theft, robbery or homicide. The judges were reluctant to impose the penalty, however, and after a lengthy lobbying effort, they succeeded in having the amendment repealed in 1965. As for Nigeria, the whip was henceforth viewed as an anachronism compared with the contemporary methods for punishing adults. For youth, on the other hand, flogging remained, with one exception, the penalty of choice in children's courts, police cells and the various institutions for the protection of white and African children.[37] At the same time, the use of flogging was institutionalised in the townships. The authorisation to punish by whipping was granted to numerous organisations by local admin-istrators or the Ministry of Native Affairs throughout the apartheid period (parent associations, vigilantes, and informal courts of justice from Sharpe-ville to Soweto and the coloured townships in Cape Town (Chapter 3).

Yet the defeat of penal reform in South Africa made itself felt most acutely in the increasingly systematic incarceration of children and youth. From the 1950s to the 1970s, its implementation had a direct effect of bring-ing about their release from prison in Western Europe and the United States (Muncie 1999, pp. 275–284). In South Africa, the reform movement initially led to a perceptible decline in the number of juveniles and children behind bars (from 10% to 4 %), but the nationalists reversed this trend: in 20 years, the percentage rose from 4% to 21% of the prison population (Table 2.3).

Gary Kynoch (2008)convincingly argues that the high level of incar-ceration of the African population – which was already exceptional on the African continent at the end of the 1950s – was in large part responsible for the large-scale development and dissemination of violence in South Africa. Although a far more systematic study is needed, we suggest here that the ra-cialisation of youth policies, combined with massive incarceration, created a prison environment in which children who passed through the welfare system underwent violent criminal socialisation. As early as the 1930s, most prisons in South Africa had gangs from the Ninevite tradition, who operated in the region of Witwatersrand in the early twentieth century (Steinberg 2004, p. 7; Van Onselen 1984). According to Jonny Steinberg (2004, pp. 120–131), the Cape Town gangs that emerged in the 1970s came into being within the South African penitentiary system. Everywhere minors who went to prison

TABLE 2.3 Children and youth under the age of 21 in South African prisons (1929–1973)[38]

Year	Total Number	Percentage
1929	17,928	10.3
1937	16,175	8.2
1944	10,901	5
1948	9,884	4
1949	11,127	4.8
1962	47,472	13.6
1969	89,198	17.9
1973	73,132	21.3

Source: Data from Annual Reports of Prison, South Africa, 1929–1973.

seemed particularly vulnerable to the violence of older inmates and gang leaders. It is therefore quite possible that the core of the prison system and of child assistance (prison, reformatories, remand home for minors) was in fact a process in which African and coloured children experienced violent socialisation at the hands of their peers and gangs, a process from which most white children were spared because they were kept separate from hardened criminals.

The individual histories of children who spent time in South African penal and educational institutions in the 1960s and 1970s reveal different trajectories depending on the racial group to which they belonged. In the early 1970s, a white boy from the Natal was sent to an orphanage at a very young age, and then to several places of safety, which propelled him towards a criminal career, but after going through a reformatory and then an approved school, he finally found what he called 'a home' where he was safe from the harassment by gangsters encountered in police cells (McLachlan 1986, pp. 104–105). During the same period, a group of six coloured children, confined for several years in a remand home, denounced constant bullying by gangsters, sexual abuse by older detainees and staff who never followed up on their complaints (McLachlan 1986, p. 122). Sometimes this ordinary violence became intolerable: in 1975, a coloured boy from Cape Town placed in a reformatory at the age of 10 ended up hanging himself to stop the daily bullying to which he had been subjected.[39] One way of escaping peer violence was to join a prison gang that offered protection from continual exactions: despite the limited sources on this subject, it seems that the gangs also organised child protection institutions. Magadien Wantzel, one of the best-known prison gang leaders today, became familiarised with this violence at Cape Town reformatories in the 1960s and 1970s (Steinberg 2004). An orphan who originated in Soweto, was assigned to a place of safety in 1968 where the older inmates beat and sodomised the younger ones; he chose to join a gang to avoid the day-to-day violence at the facility (McLachlan 1986, pp. 123–124). Because children 'in need of care' and orphans were not separated from 'dangerous youth' and gang leaders and these two groups from countless individuals confined for violations of alcohol regulations and pass laws, it appears that the state imprisonment policy created the necessary conditions for producing and reproducing a 'criminal class' in South Africa.

Conclusion

In a wider international context, laws against child abuse in France and Great Britain were part of a lengthy state enterprise of eliminating violence to pacify social relationships, as suggested by Norbert Elias. Clearly, the colonial movement and the apartheid period eluded this enterprise. Colonial and South African authorities not only failed to pacify society; on the contrary, they resorted to various forms of coercion in the institutions responsible for reforming children and youth, while legitimising the use of violence by many local organisations that were concerned about the effects of the social and moral havoc wrought by young people. For a while, reform policy in South Africa succeeded in reducing the various forms of coercion used on children and youth, before turning into an instrument for more systematic and violent intervention in the lives of African and coloured families. Penal reform lasted a long time in both countries continuing through the 1960s. Backed by statistics, Afrikaner experts claimed African and coloured children were far more 'dangerous' than 'in need of care', and thus required essentially coercive responses, whereas the indigence and interventionism of social welfare services in Nigeria made flogging the solution most likely to keep unruly youth in line.

Thus, South African and Nigerian bureaucrats came to punish children and youth for a variety of reasons: because they were too poor; because they migrated to the city instead of staying in the countryside; because they worked in the street or loitered about; because they committed one of a whole series of minor offences against current racial or commercial regulations; and in a very small minority of cases, because they had become real gangsters. Here, the trajectories of the late colonial state of Nigeria and the racialised welfare state of South Africa seem to have radically diverged from the path of the welfare state in Western Europe, because the more the Ministry of Social Welfare became established on a local scale, the more behaviour and lifestyles were criminalised. In the colonial context and under apartheid, the penal state seems to have accompanied rather than succeeded the welfare state.

The increase in the number of delinquents and children 'in need of care' appears to have been related less to urbanisation and modern life than to the identification of a specific group and criminalisation of its behaviour, followed by a higher rate of incrimination due to the bureaucratic expansion of social services, the police and the courts. If the initial intention of the social services was to individualise punishment and treatment, in the case of Nigeria, the bureaucratisation of welfare services turned delinquents into new unitised subjects (useful for drafting administrative reports and obtaining tax money indispensable to social service operations) far more than it reformed individuals or helped them to adjust. South Africa followed the same trajectory, except in three areas: the implementation of individualised punishments became a privilege reserved for the white minority; delinquents were put to work more often during the first period of apartheid (1950–1960); and finally, the bureaucratic network had much tighter control in South Africa (more than 170,000 children and youth went before the courts every year) than in Nigeria (where it remained quite lax variably implemented across the country, and targeted only a few thousand children and youth a year).

In the end, the effects of policies on youth and children were the opposite of the ones sought by reformers: the governments transformed young people who were either not wage-earners or not in school into subaltern groups as well as sources of violence by criminalising everyday practices, exposing the poorest individuals to violent socialisation in state institutions, institutionalising corporal punishment as a method of reforming individual behaviour, and relying on the support of social groups bent on moralising and disciplining youth. Unlike the radical historians who thought gangsterism and violence were the result of colonialism or manifestations of resistance to proletarianisation, we are arguing here that colonial metropolises offered a space for experimenting *dispositifs* of exclusion to the city: those devices were not only a means of racial classification, they also generated categories (juvenile delinquents, isolated women, youth coming from the countryside, children in need of care, juvenile female street traders), which the actors appropriated because they coincided with particular interests linked to their class, age group, gender, place of residence or origin. The institutional manufacturing of violence and exclusion does not entirely account for the everyday violence in neighbourhoods or the sociological contours of the groups and individuals seeking to curb it. Let's now focus on the neighbourhoods of colonial metropolises where a specific type of policing developed, which was intended to ensure – by force if necessary – the protection of populations.

Notes

1. These definitions were laid down by law in 1937 in South Africa and in 1943 in Nigeria. A child was defined as being under the age of 14 in Nigeria and under 19 in South Africa; a juvenile was between 14 and 17 in Nigeria, and between 19 and 21 in South Africa. These differences testify to the more pronounced impact of penal reform in South Africa, which tended to raise the age of minors to provide greater protection for children and youth.

2. 'Children's Act no. 31 of 1937', *Gazette Extraordinary*, 2440, 18 May 1937, p. 304.

3. Parliamentary Debate on 13 March 1937. Address by the Minister of Education, J.H. Hofmeyr, Hansard, 13 March 1937, Coll. 3, 428.

4. School became compulsory for white children in 1907 and free of charge starting in 1920, whereas it was neither free nor mandatory for Africans. In 1932, only one-fifth of African children between the ages of six and 16 were enrolled in school.

5. Union of South Africa, Report of the Interdepartmental Committee on Destitute, Neglected, Maladjusted and Delinquent Children and Young Persons, 1934–1937, Pretoria, Government Printer, 1938.

6. Report of the Union Department of Education, 1946–1947.

7. The first contacts between Cyril Burt and Ernst G. Malherbe dated from the 1920s, when the former became an international consultant on a project to standardise intelligence scales for Afrikaans-speaking children in the Western Cape Province.

8. The first comparative intelligence tests of white and African children were conducted in 1939 by researchers financed by the Ministry of Education's Research Office, which concluded that African children were intellectually inferior. The tests were disputed, and their use restricted within the South African educational system. They were nevertheless adopted at the country's only coloured approved school starting in 1948 (Dubow 2001, p. 124).

9. Jan Hofmyer founded the first training school for African social workers in South Africa in 1941 (Jan H. Hofmeyr School of Social Work) in which Ellen Hellman taught sociology. Concerning the relationship between Alan Paton and this reformist milieu (see Alexander 2009).

10. Provincial Archives Calabar, Calprof, 7/1/618, 'A Survey of the Activities of the Welfare Officer in Calabar During 1945, 23 February 1946'. Calprof, 7/1/618, 'Welfare in Calabar by the Social Welfare Officer, 5 October 1945'. National Archives Enugu, Provincial Office, Onitsha, 12/1/2089, 'Juvenile Delinquency, Onitsha by the Welfare Officer, 22 October 1949'.

11. NAI, Comcol 2600, Paterson, 'Crime and its Treatment in the Colony and Protectorate, March 1944'.

12. South African Republic, *Statistics of Offences and of Penal Institutions, 1949–1962*, Pretoria, Government Printer, 1963.

13. *Report of Penal and Prison Reform Commission*, Pretoria, Union of South Africa, 1947.

14. In other words, 1,535 African minors out of a total of 60,085 went before the courts in 1945.

15. The percentage of white children enrolled in secondary schools rose from 50% in the 1940s to 90% in the 1970s, compared with 16% of African children at the same date.

16. Among these signs, the South African Penal League lost its financial and intellectual independence and the civil servants at the Ministry of Native Affairs henceforth enjoyed greater control over the departments of Non-European Affairs in municipalities.

17. In 1967, 33 psychologists were appointed at white reformatories and approved schools; they represented half of the ministry's 'school psychologists'. Annual Report of the Ministry of Education, 1967.

18. NAI, 36005/1, Statement from Prince Eikineh to headquarters CID station Eastern Province, Gold Coast Police, 23 November 1939.

19. The Criminal Code of the Gold Coast was amended in December 1942. The new code provided for two-year prison sentences for anyone who encouraged the prostitution of a girl under the age of 13, and 50-pound fines or six months in prison for male or female pimps and brothel managers, tenants and brothel owners, 'Amendments to the Criminal Code of the Gold Coast to Suppress Traffic in Women and Children for Immoral purposes', December 1942. Criminal Code (Amendment) Ordinance 1944. The Penal Code of Nigeria was amended in 1944. It punished anyone responsible for a child or young girl who authorises that the child or young girl reside in or frequent a brothel.

20. CO 554/624 Colonial Social Welfare Advisory Committee, 'Report on a Tour of British West African Territories by the Social Welfare Adviser', 7 December 1949.

21. Comcol 1, 2844, Memorandum from the Colony Welfare Officer, 7 August 1943; Comcol 1, 2844, Calabar Council Office to District Officer Calabar, 17 January 1944.

22. National Archives Ibadan (NAI), COMCOL 1, 2844, Dwight E. Faulkner, 'Report on Child Prostitution in Lagos', 1 July 1943.

23. This hypothesis also made by Georges (2014: 152–153) has yet to be confirmed by other studies.

24. I would like to thank Saheed Aderinto for sharing this information.

25. Today, the debt bondage of Nigerian prostitutes is one of the main characteristics of this transnational traffic (Taliani, 2012).

26. PRO, CO 554/624, Colonial Social Welfare Advisory Committee. Report on a Tour of British West African Territories by the Social Welfare Adviser, 7 December 1949. Extract from minutes of the 40th meeting of the Colonial Social Welfare Advisory Committee held on 12 December 1949. Social Welfare Planning in West Africa.

27. NAK, MSCW, 1226, Report by Mr. Chinn, Social Welfare Adviser to the Secretary of State, February 1950.

28. National Archives Ibadan (NAI), Comcol 1, 2600, Lagos Federation of Boy's and Girls' Clubs, Federal Nigeria Society for the Blind, 1960.

29. Annual Report, Federal Ministry of Labour, Social Welfare Division 1961–1962, Lagos, Federal Government Printer, 1963.

30. NAK, MSCW, 1226, Report by Mr. Chinn, Social Welfare Adviser to the Secretary of State, February 1950.

31. Annual Report, Federal Ministry of Labour, 1963, op. cit.

32. The authorities in Dar es Salaam deported up to 2,000 people a year (so-called 'unemployed' boys and girls) during the 1950s, compared with only a few dozen between the wars (Burton 2005, pp. 249–256).

33. National Archives Enugu, NAE, Annual Report of the Social Welfare Department, Northern Region of Nigeria, 1955–56. Foreword by Abba Habib, Minister of Social Welfare, Co-operatives and Surveys, 20 August, 1956.

34. NAE PR.X6, Eastern Region of Nigeria, Policy for Welfare, 1955.

35. NAE, Annual Report of the Social Welfare Department, Northern Region of Nigeria, 1955–5.

36. The first camp was set up in Elandsdoorn (Eastern Transvaal) in 1954, and subsequently extended to the entire country: seven camps were in operation for African minors in 1961, and four were opened for coloured minors during the 1960s (Cook 1982).

37. The exception concerned the approved schools for white minor girls, where corporal punishment was no longer considered appropriate starting in the mid-1970s (Midgley 1982, p. 107).

38. Annual Reports of Prison, South Africa, 1929–1973.

39. Interview with the victim's sister, Manenberg, Cape Town, March 2009.

PART II

Policing the Neighbourhood

'During the war, when the police force was reduced, the government set up an organisation called the Civilian Protection Service (CPS) to carry out routine policing in Johannesburg. Natives were enrolled for this purpose in native urban areas. When the CPS was dissolved in 1946, the natives in the townships considered it so effective that they continued to perform the service voluntarily. Since then, the CPS has been maintained with the invaluable help of the police and has achieved excellent results'.[1]

'I have been asked to tell you that the enclosed instructions concerning the hunter guards are satisfactory, but I would like to add the following recommendations, which I consider particularly suitable for Ibadan and its villages: 1) Village chiefs must supply the hunter guards for their villages and teach them to be night guards; 2) It is imperative that the neighbourhood chiefs be responsible for recruiting the hunter guards for their areas and obtaining gunpowder for their guns; 3) Starting at midnight, everyone must observe the curfew in their homes; anyone found loitering in the streets will be arrested by the hunter guards'.[2]

Part 1 considered how metropolises were conceived by public authorities as the setting for penal reform, which, in the 1930s was proving difficult to implement. At the same time, low-income neighbourhoods and townships became the setting for initiatives by numerous organisations to control youth perceived as dangerous. Young people engaging in extortion in organised groups (known as *tsotsis* in Johannesburg, *skollies* in Cape Town, *jaguda boys* and *boma boys* in Lagos and Ibadan) were a source of increasing concern in the neighbourhood where they operated, and the press continually called for a serious crackdown on their behaviour. It was clearly no coincidence that the state apparatus and local organisations both developed at that very same moment a fear of unruly youths and a feeling of insecurity about their growing menace to goods and persons.[3] This issue was also very worrying for prominent figures, who were anxious to find a solution to what they saw as a threat of rising criminality.

Over time as state capacities, the South African police intervened regularly and harshly in the townships, carrying out raids against alcohol production and consumption, fastidious pass inspections and heightened surveillance of political organisations especially from the 1960s. In South Africa, the memory of these actions was to become part of the traumatic legacy of apartheid, whereas in Nigeria the police took brutal measures as well, but they were less systematic even if this needs to be further explored as the police tended to present inaccurate assessments of violence (Aderinto 2018, pp. 165–166)[4]. On the other hand, the police forces in both countries, like those in most colonies on the continent, were either passive or absent in the face of delinquency and common law crimes: in rural and urban areas, the methods of social control were expected to keep in check offences against property and persons (Blanchard 2012, p. 5). Thus a strong feeling of insecurity, combined with an institutional police force that failed to combat property-related delinquency and interpersonal violence, produced analogous effects in South Africa and Nigeria: an increasing number of organisations sprang up to protect the local population in environments where the threat to physical or material security was grasped early on.

The CPS in Johannesburg like the hunter guards in Ibadan (highlighted in the above citations), were two organisations among others set up during the Second World War to fight property-related delinquency. Since then, not a single decade has gone by without one of the following organisations operating in South Africa or Nigeria: the people's police, *ilizo lomzi* or 'eye of the home', civic guards, peacemakers, the vigilante league, street and area committees, people's court, neighbourhood watches or community police in South Africa; hunter guards, night guards, 'keeping watch' (*sode sode*), guards (*ode*), vigilantes or ethnic militias in Nigeria. From the outset, these groups attested to the weakness of state monopoly on coercion and the considerable control exercised by certain groups and individuals over police functions in many neighbourhoods in Cape Town, Johannesburg, Lagos and Ibadan. Their routine operations revealed diverse forms of protection and methods for excluding the so-called 'dangerous' populations in these neighbourhoods. Did they classify or exclude individuals or specific groups they saw as a threat to the 'community' they intended to build? To what extent did the targets identified by these organisations help regulate or normalise everyday violence? Did they play a role in delimiting a particular territory – in this case, a neighbourhood – that these groups helped bring into being? And eventually did the functioning of these local organisations call into question the authority and sovereignty of the state?

A distinction must be made between law enforcement carried out by the 'police' – the name commonly used to designate a state organisation with a specific mandate – and 'policing' that designates a plurality of organisations including the police (Garland 2001; Jobard and Maillard 2015, p. 225).

The presence of vigilante groups in policing operations in many countries led to the development of a whole current of social science research that came under the name 'vigilantism' (Fourchard 2018). According to the definition proposed by the British criminologist Les Johnston, one of the first researchers to analyse this phenomenon, vigilantism is 'a social movement made up of voluntary private agents who engage in premeditated action in the face of perceived transgressions of institutional norms to control criminality and ensure the security of their members by using or threatening to use violence' (Johnston 1996, p. 220). Johnston distinguishes between two types of voluntary activity: the first is recognised by the state, and the second is not (Johnston 1996). The latter is what he calls 'vigilantism'. Researchers working in contexts where the police had little legitimacy and the state was far from having a monopoly on legitimate violence criticised Johnston for not taking sufficiently into account the fact that public authorities may accommodate such practices and that the police themselves sometimes resort to extra-legal violence (Kirsch and Grätz 2010, pp. 5–6). His definition nevertheless has the virtue of raising the central question of the relationship between vigilante groups on the one hand, and state agents belonging to the police, army or administration on the other. In subsequent research on this question, anthropologists have suggested looking at vigilantism in terms of its relationship to the state. For Ray Abrahams, in particular, 'vigilantism implies an organised attempt by a group of ordinary citizens to enforce the law and apply norms on behalf of their communities, often through violence, in the perceived absence of state action by the police or the justice system' (Abrahams 2007, p. 423). The author acknowledges, however, that this definition is an ideal type, which, from a Weberian perspective, authorises de facto considerable empirical variations.

David Pratten and Atreyee Sen (2007) concur with Ray Abrahams in considering 'vigilantism' a labile limiting concept, but instead of embracing his definition, they use vigilantism as an umbrella term to encompass very different types of organisations: vigilantes, militias, protection gangs, paramilitaries, community policing, and practices of 'spontaneous' justice such as lynching. The absence of a stable definition leaves open questions such as how much organisation is required to qualify as 'vigilantism' (does it include lynching by a mob?); its relationship to violence (does it necessarily engage in violence?); and its relationship to the state (is a perceived lack of effective police or justice indispensable for vigilantism to arise?). Moreover, American researchers initially included so many movements under the term 'vigilantism' that the concept became meaningless (Rosenbaum and Sederberg 1974). No doubt it will be useful to specify what types of action it covers, so we can avoid falling into the same trap. The definition proposed by Gilles Favarel-Garrigues and Laurent Gayer provides a clearer picture: 'Vigilantism covers a range of often violent and usually illegal collective actions carried out by non-state actors whose professed vocation is to keep order and/or dispense justice, in the name of legal or moral norms (Favarel-Garrigues and Gayer 2016, p. 9). Based on this preliminary approach, three imperatives appear essential to this section.

The first imperative is to explore the concrete relationship between these organisations and the state. Vigilantism is not just another form of neoliberal government (Pratten and Sen 2007) contrary to what some authors have suggested (Comaroff and Comaroff 2006; Goldstein 2005) nor can it be reduced to the supposed privatisation, failure or decline of the state.[5] To understand the relationship between vigilante groups and the state, we must observe the practices of the group members and police officers, and see them as part of a historical and sociological continuum, in other words, conceive of the social practices of law enforcement as processes constantly being reconfigured, criticised and subject to negotiation between (sometimes divided) public actors and private interests (Favarel-Garrigues and Gayer 2016, p. 28).

Vigilante groups are typical organisations exercising a form of public authority. Authority means in a Weberian sense, an 'instance of power which seeks at least a minimum of voluntary compliance and thus is legitimated in some way' (Lund 2006, p. 678). If we analyse the contingent relationship of these organisations to state representatives, the first necessary step is to uncover how organisations exercising public authority actually work rather than departing from a theory of the state: this includes accounting for how power is legitimated and practiced and how claims to public authority are connected to the provision of public goods like security and justice (Hoffmann and Kirk 2013). Vigilante organisations are for Christian Lund 'twilight institutions': they are situated at the interface between the state and the populations in a continually fluctuating position, without any predetermined direction, which rules out the temptation to assign them an ultimate purpose (Lund 2006). The point is to comprehend them without postulating their outcome a priori, i.e. we need not assume the state will be strengthened by these organisations (e.g. by absorbing them, which could happen), or that state authority will be weakened or threatened (due to the increased autonomy of the groups, which could also happen). The processes testify to a way of negotiating state power (Péclard and Hagmann 2010), and the negotiations enable us to think about the scope of the state's action and its determination to control (or not control) the groups and tolerate (or not tolerate) their use of violence. By analysing them over the medium term, we can thus question the legitimacy of vigilantism in radically different historical contexts.

Is the violence manifested by these organisations identical when it is authorised (during the colonial period and apartheid) and when it is prohibited (today)? Are these organisations regulated in the same way when bureaucratic resources are meagre (the colonial period) and when police forces have been significantly expanded (today)? And do the targets focused by these organisations differ over time?

To grasp the connections between the historical periods, the chapters that follow draw on similarly substantial historical and ethnographic materials concerning vigilantism from the past and present. The historical sources come from the government, colonial and local archives consulted in Ibadan, Pretoria, Cape Town and Johannesburg as well as several oral interviews of elders and articles from the local press.[6] The contemporary materials include tens of interviews with policing actors (local leaders, foot patrollers, police officers) and participatory observation of several patrols in Cape Town in 2009 and 2010. The amount and type of materials varies: there are more documents available on South Africa than on Nigeria because archives there were not preserved after 1960, whereas participatory observation in South Africa provided richer material than only in-depth interviews in Nigeria. Part Two is divided into two chapters: Chapter 3 on the colonial and apartheid period, and Chapter 4 on the period after the 1970s. We will retrace the histories of particular vigilante organisations, which enjoyed a limited lifespan (from a few years to one or two decades), to set this short-term period against the longer historical backdrop of the twentieth century.

Despite the richness of the anthropology of vigilantism in African studies and its pioneering character on a global scale, the urban environment in which these groups operate often appears, as a secondary or not very determinant. Researchers have looked at their different forms of exercise of public authority, the building of 'moral communities', the performances of violence and the threats identified by vigilante groups, but less so on how the everyday working of these groups remake ordinary urban experiences and govern urban space. The scale of the neighbourhood will allow us to see the changes and continuities in everyday policing (how law enforcement was carried out concretely, how violence occurred, the groups targeted by volunteer vigilantes). The coloured townships of Cape Town and the neighbourhoods in the historic centre of Ibadan are the main focus of the investigation. These areas have all been sites of regular vigilante activity since the colonial and apartheid period, but the public authorities have intervened within them in radically different ways. The Cape Flats in Cape Town are the product of a project of racial and social control of populations decided by the South African government as early as the 1950s, while the historic districts of Ibadan were not the subject of any state intervention throughout the twentieth century. This unequal interventionism of the state in urban transformations plays its part in the practical modus operandi of vigilante groups.

We are seeking in particular to understand how these organisations govern the neighbourhood in the sense that they exercise power over the people residing there and set about ensuring their safety. They occupy an ambivalent position as intermediaries between those 'from below' and those 'at the top', (Bayart, Mbembe and Toulabor 1992) a situation that explains their subordination to the state or local administration, even though they may have ambitions of their own, especially a desire to play a role in the power structure. The term '*Eigensinn*', developed by Alf Lüdtke (2000, 2015a), accounts for the ambivalence of these actors better than the term 'agency' by revealing the appropriation and mundane uses of structures of domination. The word *Eigensinn* refers to the multiple ways individuals relate to power, from resistance to hyper-obedience and from the desire to do better than what is required to the will to participate in power. Vigilante action is part of this fundamental ambivalence. On the one hand, it means 'breaking the law to make people comply with it, committing offences to combat other offences' (Favarel-Garrigues and Gayer 2016, p. 23), and on the other, implementing the law by taking on the enforcement role the police are unable to assume. In both cases, resorting to coercion (or threatening to do so) helps to produce a community defined essentially in opposition to certain groups that threaten its cohesiveness.

The third imperative is to shed light on areas overlooked by the anthropology of vigilantism and by urban studies oriented to neoliberal imported models, especially the politicisation, bureaucratisation and commodification of the organisations. For various reasons, the politicisation and de-politicisation of the leadership and members have seldom been analysed, despite the fact that political parties have at times succeeded in transforming the strong demand for security into a political resource. Similarly, while vigilantism is known to be 'a cheap form of law enforcement' (Pratten and Sen 2007), there has been little discussion in the literature as to whether it is volunteer work performed free of charge, compensated in some form, or wage labour. Based on our medium-term analysis, we can assert here that vigilantism has not attained the same degree of commodification in the two countries. Finally, we will study how security becomes bureaucratised in poor neighbourhoods, a topic virtually overlooked in the recent focus on the forms of neoliberal governance of security, which implies, according to its detractors the retreat of state forces, the multiplication of public–private partnerships, and more systematic use of private companies and community groups to provide security.

Notes

1. National Archives of South Africa, NASA, BAO 520/400 (12), Native Commissioner to Chief Native Commissioner, Johannesburg, 1 September 1951.
2. National Archives Ibadan, NAI, Ibadandiv, 1/1, 167, Hunter guard instructions from the District Officer to the Olubadan-in-Council, Ibadan, 5 January 1942.
3. The first urban riots, 'criminality', and city-related ills (vagrancy, floating populations, detribalisation) became a new concern in many parts of French and British Africa and South Africa (Cooper 1995; Glasman 2014).
4. See the episodes of brutal repression of demonstrations or strikes, e.g. the massacre of miners in Enugu in 1949 (Brown 2003).
5. The term 'privatisation' has been called into question in Africa, Europe and North America. (Hibou 1999a; King and Le Galès 2011, p. 467).
6. In Nigeria, the *West African Pilot* from 1937 to 1960, the *Nigerian Daily Times* from 1930 to 2000, and the *Nigerian Tribune* from 1948 to 2000; in South Africa, mainly the *World* from 1950 to 1970, the *Cape Times* from the 1960s and 1970s and the *Plainsman* from the 1980s to 2010.

CHAPTER 3

Vigilantism and Violence Under Colonialism and Apartheid

The long-standing existence of vigilante organisations in South Africa attests to the basic insecurity experienced by African populations amidst omnipresent violence in mines and townships following the country's mining revolution in the late nineteenth century. Putting aside the differences in the social situations and amount of criminal violence in the two countries, a comparative analysis of their respective vigilante organisations will allow us to determine what was specific to each one and what they had in common, how they operated, the threats they discerned and their relations with the police and with the state in general.

To reconstruct the genealogy of these practices, we have to examine the different time frames that shaped the way vigilantism was experienced by those involved in protection. From the vantage point of the state, this history oscillated between authorising, absorbing and banning the organisations. From the perspective of the organisations themselves, it often appears to have been cumulative. Vigilante organisations developed in accordance with dynamics already embedded in a number of particular histories, which cannot be reduced to a simple duality of legalisation/repression by the state. The groups were frequently rooted in past experiences and their values derived from the history of their immediate environment in Nigeria and South Africa. While the colonial authorities gave a number of organisations the right to regulate their areas and delegated the authority to resort to violence, the organisations used that violence to achieve their own aim, i.e. to produce a moral community that designated its most important enemies, and thereby helped to create ontological categories of enemies threatening the cohesion of the imaginary community then under construction. Depending on the situation, these categories included: migrants; single women; youths; foreign, religious or sexual minorities; outsiders living in the neighbourhood or the village and

Classify, Exclude, Police: Urban Lives in South Africa and Nigeria, English Language First Edition. Laurent Fourchard.
© 2021 John Wiley & Sons Ltd. Published 2021 by John Wiley & Sons Ltd.

criminal organisations. For organisation leaders (neighbourhood chiefs, heads of lineage, prominent figures, parents, association heads, etc.), such menacing groups offered opportunities to assert or reassert their authority.

'Threats' were manufactured in different ways in South Africa and Nigeria. The civilian groups organised by the elders in South African townships tended to see children and youth as the most dangerous threat: young people needed to be protected against their own violence and against the temptation to consume alcohol or take drugs, depending on the period, and systematically taught to respect adults and the world of work in general. The 'threat' went by a generic name: in African neighbourhoods it was the *tsotsi* and in coloured neighbourhoods the *skollie*, referring to groups of young boys whose scope of activities varied considerably, since the same names were used to designate everything from loosely connected bands of local youths to structured organisations, each with its own leader, hierarchy, code and 'warrior capital'.[1] In Nigeria, on the other hand, armed burglars and robbers, usually carrying knives or locally produced firearms as well as charms, were seen as the main threat. As the 'enemies' were mobile, not attached to any specific territory or neighbourhood, they were very often viewed as outsiders. Because threats were manufactured differently in the two environments, different forms of vigilantism arose in response to them. Two specific areas – the historic centre of Ibadan and the Cape Flats in Cape Town – reveal major differences between the countries in terms of police routine, between forms of the built environment, between groups of population targeted for control and exclusion and between forms of violence used by organisations. This comparative history takes forward a new approach to the historiography of law enforcement: as routine policing in low income urban areas was neglected by institutional police, neighbourhood-based organisations were left with the power to define good and bad behaviour, to decide who can and who cannot enjoy a full access to the neighbourhood at any time and more generally to govern everyday life especially during the night.

Policing in a Colonial Situation: Historiographical Detours

The historiography of colonial law enforcement is a poor relation of the revival of imperial studies: the topic is considered too broad (law enforcement pertains to the entire colonial project and implies a multitude of actors) and too disparate (the apparatus was extremely varied: specific judicial regime, prison, work camps, urban control, criminality) to allow for systematised theorisation (Blanchard and Glasman 2012, pp. 11–13).

If law enforcement is to be understood as the state laying claim to a monopoly of physical violence in a given territory, in colonial situations, the state did not insist on such a monopoly but rather oriented coercion, whatever its source, in a direction that did not threaten the stability of the colonial government (Blanchard and Glasman 2012, pp. 11–13). The colonial state operated through the use of 'discharge', a Weberian term proposed by Béatrice Hibou to convey a way of exercising power characteristic of non-bureaucratised societies, which involves little or no reliance on an apparatus to avoid the cost of considerable administrative logistics (Hibou 1999b, p. 8). Unlike 'delegation', the term 'discharge' suggests that state functions may be assumed by private actors without receiving any order or directive to that effect, in the case of law enforcement, from the police (Hibou 2017). Vigilante groups operating in the centre of Ibadan and the Cape Flats in Cape Town are a manifestation of this 'discharge' of colonial police.

Like the new global history of the empire, the history of policing examines how knowledge, people and police practices circulated within – and between – imperial spaces, and calls into question the idea that metropolitan models were uniformly imposed on colonial territories (Anderson and Killingray 1991; Thomas 2007). One of the advantages of this approach is its ability to provide accurate information about the amount and limits of the forms of coercion used in the colonies as well as the colonial authority's need to delegate those functions to a variety of local authorities – colonial companies, settlers, natives – including in the areas of law enforcement, justice and prison (Alexander and Kynoch 2011; Bernault 1999; Mann 2009). Unlike older views of imperial history, this new history focuses on the historicity of colonised societies and their ability to reformat policing models conceived elsewhere. The blending of the pre-colonial with the colonial is particularly significant in analysing how imported institutions were grafted onto pre-existing institutions (Blanchard, Deluermoz and Glasman 2011). Recently, a specialist with considerable expertise on colonial policing was able to assert that the deeper one goes into the history of colonial policing, the more difficult it becomes to establish the boundary line between regular police forces, private security companies, native police and vigilante groups (whether European or native) (Thomas 2013). Anthropological and historical studies on vigilantism confirm that it is equally difficult to make a clear-cut distinction between society and state with regard to organisations (Buur and Jensen 2004; Kirsch and Gratz 2010; Pratten and Sen 2007). However, the historicity of these groups is still not thoroughly understood and the chronology has yet to be established, including in South Africa and Nigeria, where they were essential elements of the apparatus of power (Fourchard 2008;

Pratten 2007; Seekings 2000a). Coming back to the different histories of urban vigilante organisations helps reconsidering this chronology.

These remarks in turn raise a central issue for research on colonial policing, a subject on which there was once relative agreement: this is the widely held view that colonial and apartheid police were coercive and brutal, made no attempt to win the consent of the colonised populations, and hence were ontologically different from police in democratic countries (Anderson and Killingray 1991, p. 9; Brewer 1994, p. 6). Recent studies confirm that colonial violence was not restricted to the period of conquest, which can't be opposed to a later period of bureaucratic construction. Martin Thomas focuses on colonialism's political economy, emphasising the priority given by the British, French and Belgian empires to defend colonial economic interests: the police forces were thus at the heart of numerous forms of coercion and brutal repression aimed at putting the population to work, breaking strikes and quelling demands for improved living or housing conditions (Cooper 1995, pp. 87–145; Thomas 2007). After the Second World War, the use of army and police intelligence services to combat nationalism, communism and requests for imperial citizenship were no less violent (Anderson and Killingray 1992; Bat 2012), not to mention the extreme repression exercised by the colonial armies and their auxiliary troops during decolonisation, particularly through camps and forced relocation, which in some instances were exacerbated forms of routine colonial internment (Anderson 2005; Branch 2009; Branche, 2001; Deltombe, Domergue and Tatsitsa 2011; Thénault 2012). Despite the adoption of new bureaucratic expertise by post-war police, the treatment of the population was no less brutal, as shown in the cases of Togo and apartheid South Africa, where the police force comprised one fourth of all civil servants, making it the state's largest department (Glasman 2012, pp. 37–54; Shear 2012, p. 177). In the face of colonial nostalgia, insisting on the amount of violence used by law enforcement and armies has increasingly become a scientific imperative (Bertrand 2006b). To explore this relationship more fully, then, we need to investigate the actions of everyday policing based on arrangements and 'collusive transactions' with the population (Bayart 2007; Blanchard and Glasman 2012; Lonsdale 1981).

Indian historiography has a head start on this topic. In his pioneering work on Madras, David Arnold emphasised the role of the colonial police and their intermediaries in exercising coercion. Initially, the police were involved in rural law enforcement, but starting in 1900, they were increasingly mobilised to cope with 'urban disorder' and became more and more repressive and militarised. Local employees were socialised by the colonial state through recruitment, training, residence, discipline, ethos and duty until they were believed to be systematically alienated from Indian society (Arnold 1986).

More recent studies have put greater emphasis on the fragile nature of colonial police control, its inability to meet the security requirements of the population, and the need to rely on local allies in the countryside as well as in the city (Bloembergen 2012; Chandavarkar 1998; Gooptu 2001). Thus, the police did not operate merely as an instrument of urban social control, but had frequently to call upon other operators who were relatively independent of their command structure, while workers could simultaneously make use of personal, social and caste ties to seek to influence police officers. In colonial India, the pendulum swung between a police force symbolising an all-powerful state and a revisionist perception that underscored the weakness of its control and its embeddedness in local social relationships (Kidambi 2004, p. 2 and pp. 27–27). This debate takes us beyond law enforcement by the police to take into account the important role of intermediaries including vigilante organisations in the process of social regulation.

It is far from obvious however that vigilantism was in the service of a colonial political economy even as the devices for law enforcement served first and foremost to defend the interests of the colonial economy (Thomas 2013). For example, the neighbourhoods we will discuss in detail in the following sections were outlying areas in which the residents' well-being and security had little effect on South Africa's mining and industrial economy or Nigeria's agricultural economy. The administration could not function without intermediaries in these areas, even in South Africa where the state enjoyed the widest bureaucratic reach on the continent. Consequently, the history of coercion in colonial situations is also the history of these groups. It therefore seems appropriate to shift our gaze from the police forces to these 'native' or local forces, and take into consideration the coercion exercised by these organisations, regardless of whether or not residents consented to being policed in this manner. It is especially necessary to analyse everyday policing in urban areas neglected by the formal institutions as this history has been long occluded by the historiographical focus on situations of extreme colonial violence. This history revels in everyday forms of exclusion, classification and violence operated by urban vigilante organisations that have enjoyed throughout the colonial and the apartheid periods a large autonomy in their daily work.

Violence and Vigilantism in South African Townships

By 1960, violence in townships and on the mines is said to have reached unusually high proportions compared with other urban areas in Africa (Kynoch 2008). According to many researchers, this was due to the violence of the management used in the mining industry, the government's repressive attitude, endemic violence in a society dominated by white settlers and the highest incarceration rates on the continent (Alexander and Kynoch 2011; Kynoch 2011; Moodie 2005). However, and interesting for your analysis here, along with other root causes such as pauperisation, the divisions created between Africans by white rule, and inadequate police coverage leaving territorial areas in the hands of the most ruthless individuals, William Beinart suggested exploring the forms of violence associated with manliness adopted by rural and urban organisations (male camaraderie, martial arts-inspired dance associations, sports clubs and games) and the enduring ties between members of these groups after they turned into criminal organisations (*Amalaita, Isitshozi, Ninevite*) (Beinart 1992). Following Beinart's work analysed the influence of manliness and urban subculture among the *tsotsis* in Johannesburg and the *skollies* in Cape Town (Glaser 2000; Salo no date). These groups offered their members camaraderie, security, and status: they were formed either to provide protection against other gangs or to practise violent extortion against the wage earners and gain control over township women. From the 1930s onwards, they became one of the main threats identified by adults.

Today, this new research enables contemporary readers to understand the relationship between the populations and the police, which was considerably more ambivalent than the picture presented by initial research on law enforcement. Gary Kynoch reminds us that, for many residents, the fear of crime was greater than their resentment of a police force that imposed unpopular segregation laws. It is also important to note that before the Soweto revolt in 1976, many residents had agreed to work with the existing judicial, police and educational system to try and improve their daily life (Kynoch 2003). Also, vigilante initiatives were influenced by the outcry against *tsotsis* violence in the townships explaining the constant demand for tougher measures to crack down on what was called criminal or amoral behaviour. In light of this new historiography of violence and gangs, it is important to re-examine the history of vigilitantism in South Africa, which show that the relationship between policing and population was far more ambivalent than its portrayal by the radical historiography of the previous period.

A common feature of all South African vigilante organisations over the past half-century has been their relationship to a weekly time frame: they operated mainly on Friday and Saturday nights. This time frame was linked to the country's industrial development: miners, workers and civil servants were paid their wages every Friday afternoon, leading to predatory and violent actions at weekends, a strong dissociation between work time and leisure time and the gradual concentration of leisure activities on spending and redistributing the wages of the working class. In this regard, a significant portion of African and coloured sociability as well as gangster activity was organised around *shebeens* (informal drinking establishments) on weekend nights and the two were often connected. Both gangs and protection organisations wanted to create a convivial nocturnal public space in which they could safely consume. After the Second World War, the so-called Russians gangs in the Witwatersrand mining regions offered to take charge of Sotho miners' security and weekend organisation by providing a set of services for payment (ensuring access to *shebeens* and sex workers and a safe return from the *shebeen* to the mine; Kynoch (2005)). Prior to this arrangement, numerous groups had already been providing protection in the townships of Johannesburg and Cape Town.

The earliest initiatives appeared in the 1920s in the Johannesburg region (Bénit-Gbaffou, Fourchard and Wafer 2012; Goodhew 1993), but one of the first real experiments in vigilantism involving several townships – known as 'civic guards' – did not take place until the Second World War. Originally, they were linked to two initiatives taken by the South African government when it entered the war on the side of the Allies: the introduction of white civic guards made up of adult volunteers, which then spread to the African neighbourhoods in Johannesburg, and the creation of a Civil Protection Service (CPS) in which a group of volunteers (irrespective of racial origin) oversaw precautionary measures taken against bombing raids.[2] In the European city, the CPS was set up as a temporary body and was dismantled in 1945; on the other hand, it enjoyed immediate popularity in the townships, where it was used primarily for routine policing operations in response to a long-standing request for protection. This gap was summed up quite well by Paul Mosaka, a representative of the Orlando township: 'The government wanted natives to mobilise in support of the war effort while providing protection against wartime bombing, whereas the African residents wanted 'civic guards to stop and prevent further activities by the good-for-nothing hooligans that are everywhere in our townships'.[3] Similar civic guards worked in the Cape Town centre of District Six.

The organisation's popularity acquired during the war kept it alive long after the CPS was prohibited in 1947 and civic guards in 1952. Local authorities and ordinary citizens alike recalled 'the useful work done by the native sections of the CPS during and after the war: the guards performed a public service and there wasn't half as much crime as we have today'.[4] Indeed, in 1951, the vivid memory of its effectiveness prompted the African members of the Johannesburg Advisory Board and several leaders of the South African Communist Party (SACP) and the ANC to request, on behalf of the 487,000 Africans in Johannesburg: 'that the civic guard be officially recognised and set up in the villages (sic) where it does not yet exist, that courts presided by Africans be established in the townships, and that more power be given to the Advisory Boards to put pressure on families harbouring known criminals'.[5] Nevertheless, the supposed ties between civic guards and the leaders of the ANC and the SACP and their popularity in the townships worried the heads of the South African police and the National Party, which came to power in 1948, as they were invariably presented as a paramilitary wing of the communists (Kynoch 2005, pp. 94–105). In 1952, the government decided to dismantle and arrest any resisters because they 'forced township inhabitants to join their ranks and make mandatory contributions to the cause, discouraged residents from complaining to the police and inflicted corporal punishment on local populations (Kynoch 2005, pp. 94–105). Behind these reasons, the government was in fact fighting for political hegemony in South Africa; it feared the civic guard movement, which seemed too popular, too big, and too close to the ANC. This first experiment with civic guards did not disappear entirely. In Soweto, the civic guards survived the government ban. Due to opposition between municipalities and the Ministry of Native Affairs or between police in the field and the central police administration, as well as bureaucratic foot-dragging, civic guards were authorised to rebuild their ranks for a few months or, at best, a few years throughout the 1960s.[6]

The ban on the Communist Party (1950), the civic guards (1952), the ANC and the PAC (1960), and the social dislocation brought about by large-scale eviction policies in the 1950s created an organisational vacuum (Seekings 2000). The mobilisation of civic guards was replaced by more localised groups, supported by the apartheid government, which assigned priority to combatting the presence of illegal migrants, unemployed citizens and delinquents, whom they suspected of stirring up political unrest in the townships (Evans 1997, pp. 84–90, (Chapter 2). Multiple associations were created, invoking a reified tradition of a rural world that honoured adults and opposed the supposed degeneracy, idleness and cosmopolitan lifestyles of the *tsotsis*. Parents' Courts introduced in the townships and squatters camps of Johannesburg in the late 1950s sought to revive moral values centred on work and the respect young people owed to their elders (French 1983, p. 328; Glaser 2005). The administration also tacitly supported the Legkotla and the Magkotla organisations imported from the countryside, which were created by local personalities often sitting on the advisory boards of local governments in the early 1970s. These organisations signalled an effort to adapt to transformations of the urban landscape (marked by an exacerbation of the conflicts between migrant hostels and urban neighbourhoods, see Chapter 1) as well as the government's desire to retribalise Africans by strengthening the role of the so-called traditional or ethnic organisations (Badenhorst and Mather 1997, pp. 473–489; Glaser 2000). While the civic guards of the

1940s did not represent any particular group, the social base and leaders of the Legkotla and the Magkotla were Sotho migrants who had embraced supposedly rural conservative ideas, but who were in fact known primarily for brutal corporal punishment administered in public (Seekings 2000). A similar process occurred in Cape Town. The construction of the Cape Flats in the 1960s and the 1970s inhabited by populations that were originally living together in the centre of Cape Town was part of a modernising and racialising project of the apartheid government marked by stronger bureaucratic control of the populations (Chapter 1). Like in Soweto, aggressive and hard-line conservative vigilante groups emerged and multiplied throughout most of Cape Flats African and coloured townships and supported by white superintendents.

Violence and the Making of Township Communities in the Cape Flats

Until the 1950s, the land where the Cape Flats now stand was a vast sandy area at the eastern edge of Cape Town occupied by farmers and military fields. At the end of the 1950s, the government and the municipal authorities of Cape Town decided to construct a residential neighbourhood, for evicted coloureds and Africans who had previously been living in the central mixed neighbourhood of District Six and the outlying squatters camps of Cape Town (Chapter 1). These populations were classified and relocated in various townships, usually in flats – hence the name 'Cape Flats' or 'Flats' – according to their race (Manenberg, Bishop Lavis and Bonteheuwel for coloureds, and Nyanga (renamed *Guguletu*) and Langa for Africans) and residential rights (Guguletu for families, Langa for migrants).

A number of factors account for the intensified climate of insecurity: families were torn apart by evictions (extended family members were separately housed in remote neighbourhoods and Africans without residential rights were sent to the homelands); illicit economies that once provided a livelihood for many poor families were eliminated; new housing was overcrowded; gangs from prison and the street joined forces in coloured neighbourhoods; and the ban on alcohol was lifted in 1961. The Cape Flats model dormitory towns soon turned into areas of major endemic violence. According to a survey on 2,000 acts of aggression committed between 1968 and 1970, all adult male residents in the Cape Flats African and coloured townships were highly likely to be beaten up at least once in their lives.[7] The main areas of violence were the townships of Manenberg, Bishop Lavis, Bonteheuwel, Guguletu and Langa. At the time, alcohol and poverty were considered the chief causes of violence. In reality, it was produced above all by the apartheid system of confinement.

Street gangs soon multiplied in coloured townships. Don Pinnock, who had counted only a few groups during the 1950s, found 280 in 1982 with anywhere from 100 to 2,000 members (Pinnock 1984, p. 4). The largest ones were linked to Cape Town prison gangs whose membership was regularly replaced and increased by the authoritarian welfare approach used in approved schools for minors and the especially high rate of incarceration of the coloured population (Chapter 2). The growing role of these gangs in the social violence of the neighbourhood led in turn to the creation of vigilante organisations that sought to develop security and conviviality, which was hampered by the cramped living spaces in the Cape Flats buildings (Figures 3.1 and 3.2).

The Peacemakers, created in Manenberg township in the early 1970s, was the most popular. Initially called the 'Manenberg Resident's movements (vigilantes)', the nickname 'Peacemakers' given by neighbourhood gangsters was quickly adopted by the residents and movement members. The size and swiftness of the mobilisation indicate its initial popularity: sources reported that between 1,200 and 2,000 members were recruited within a few months for a neighbourhood with only 34,000 inhabitants.[8] The Peacemakers took over the confined spaces produced by the engineers of apartheid, particularly the three-storey blocks of flats built around a central courtyard called 'the

FIGURE 3.1 The Cape Flats in Manenberg (Cape Town) circa 1974
Source: *Living in Manenberg and Bonteheuwel. Workshop on the living conditions in some coloured townships in Cape Flats, and their implications for the incidence of deviant behaviour*, Cape Town, National Institute for Crime Prevention and the Rehabilitation of Offenders, 1975.

FIGURE 3.2 A courtyard in Manenberg between three-storey buildings (circa 1974) Source: *Living in Manenberg and Bonteheuwel. Workshop on the living conditions in some Coloured Townships on the Cape Flats, and their implications for the incidence of deviant behavior,* Le Cap, National Institute for Crime Prevention and the Rehabilitation of Offenders, 1975.

Courts' (Figure 3.2). These courtyards became the group's main operating units, with members gathering around a fire on weekend evenings armed with clubs, as much to dissuade gangs from attacking as to create a space for family sociability:

> When we were operating in the courts, we drank soup, tea, and coffee in winter, we put up lights and we had discussions and moments of relaxation around the fire where we played dominoes and other games. Everybody got acquainted and that's what really built the community.[9]

The Peacemakers frequently resorted to violence. They were quick to clash with certain gangs (the Mongrels, the Born Free Kids and the Mafias), sometimes mobilising as many as 1,000 people according to former members.[10] They were nevertheless allowed to operate by the Ministry of the Interior, and local authorities were kept regularly informed on their actions.[11] The local police seemed satisfied that this surveillance system was reducing criminality in the area. In 1975, the highly official National Institute for Crime Prevention and the Reintegration of Offenders (NICRO)[12] organised a conference on the Manenberg and Bontheuwel neighbourhoods during which the president of the Peacemakers and a police officer discussed the respective advantages of their respective forms of policing in the Flats. In 1976, a commission recommended authorising group members to use corporal punishment on juvenile delinquents (Viljoen Commission 1976, p. 37). Today, the former members claim that corporal punishment was necessary to remind the young of the respect they owed to elders.[13] The government did not change its attitude until after the Soweto riots in June 1976. When members of the Peacemakers joined the campaign against apartheid, the authorities decided to ban the organisation in 1979 (Kinnes 2000, p. 22).

The aims of vigilante organisations in the African townships of Cape Town were essentially the same. When anti-apartheid leaders went underground, they left a political vacuum that was filled by the leaders of local associations. Following the riots in March 1960, the closely monitored associations were forced to abandon their confrontational attitudes and focus instead on less politicised, more consensual topics, such as the fight against delinquency. This was the approach adopted by vigilant associations or *Ilizo Lomzi* ('the eye of the house' in the IziXhosa language). These local organisations had been operating in the cities of the Western Cape and Eastearn Cape provinces since the end of the nineteenth century. Although they were originally designed to protect the interests of African taxpayers in the areas of education, health and housing, they focused their attention almost exclusively in the 1960s on combatting juvenile delinquency.[14] The change of tone was significant, for example, in the discourse of the main residents' association in the Nyanga West township, an emergency camp set up in 1956 for evicted African families. The association had been created in 1959 to give voice to residents' complaints regarding the frequency of police raids and more generally 'about the servile, oppressive conditions in which the families of this township live'.[15] In 1961, however, the association expressed its concern about insecurity in the township and the unruliness of young people, echoing the aim of a rival township association that was more openly supported by the local administration.[16]

As in Manenberg, the associations that sought to exert social control simultaneously developed a moral code aimed at setting an example and stimulating civic pride. The Vigilante association of Nyanga West told parents to keep their children at home after 8 p.m. or risk expulsion from the township if they could not provide a satisfactory excuse for failing to do so.[17] It organised a civic guard to enforce the curfew, in charge of arresting 'youths, loiterers, and delinquents' in the township, which was divided first into eight and later twelve sections placed under surveillance.[18] In 1964, a rival organisation, the

Vigilante League of Decency of Guguletu (*Ilizo Lomzi wase Guguletu*), promulgated a constitution aimed at 'dissuading the bad behaviour of juveniles, encouraging prevention, and fighting the various forms of deviant behaviour among children and youth'.[19] Each of its members had 'to set an example, be honest and diligent, help the weak and the sick, and behave properly'.[20] Parents had to discipline their children, who were not to be allowed to loiter in the streets at night.[21]

Moral reform of the township basically required punitive education. Abie Sumbulu, a child in the Guguletu township during the 1960s, remembers being beaten on several occasions by an *Ilizo Lomzi* group on the way home at night. The violence they displayed seemed limitless at the time (he thought he survived only because he managed to escape from his assailants), but was socially tolerated (no one dreamt of lodging a formal complaint against the *Ilizo Lomzi*).[22] Tembile Ndabeni, who was born in the township in 1963, remembers the extent to which respect for the adult world meant children and youth had to comply with a set of rules designed to curb their freedom and ensure the moral and civic preservation of the township:

The ones who patrolled at night were adult men. They did it every day after work, depending on their agenda. I don't know where they found the energy to do it. We children called them the *amavolontiya*: that was how we translated 'volunteer', a word that had no other meaning for us than 'night patrol'. Most of the time, the *amavolontiya* asked the children and young people to go home. Children were not allowed to loiter in the street at night. As children, if we had no valid excuse, we were flogged. Patrols were organised to cover one or more streets. On my street, there was one member who was highly respected by the young people. If rubbish was left in the street, the next morning you had to apologise to avoid a beating. Some members had handcuffs and used them to arrest *tsotsis,* who were betting in the street or playing dice, and turn them over to the police.[23]

In the minds of the patrollers, young people and *tsotsis* – a generic term used in the townships to designate gangsters, delinquents and idlers – were grouped together into a single category. For the Guguletu League, 'deviant behaviour' encompassed a wide range of everyday practices (loitering after dark, playing dice, dirtying the streets), thereby combining the fight against offences against property and persons with moral norms targeted at youth. Indeed, it was the daily combination of moral standards with legal struggle that defined and built these moral communities in South Africa.[24]

Finally, the labour policy of the government and the West Cape province, combining the stabilisation of family life with restricted city access for African migrants, was a major catalyst for social division and violence between urbans and migrants (whose permission to reside in the city expired at the end of their work contracts) (Chapter 2). This was especially the case in Langa, which was divided into a married quarter (8,000 people, half of whom were 'children' in 1964) and an area of barracks and hostels or 'workers' hostels' (20,000 single migrants in 1964). As violence increased between the residents of these two areas, a new protection system was introduced in the South African police force. The *tsotsis* saw the migrants in barracks as '*Iibhari*' a term that implied crude, poorly dressed, unmarried, workers[25], who were unable to speak *tsotsitaal*, their identifying language and urban slang combining Xhosa, Zulu, Afrikaans and English (Glaser 2000, p. 8 and pp. 70–71). In the eyes of the *tsotsis*, they deserved to be robbed of their pay on Friday nights or goaded into fights at knifepoint when they wandered out of their neighbourhoods looking for company or a place to drink. In June 1966, the murder of two such migrants was followed by a fortnight of collective retaliatory attacks carried out in the married quarter by groups of migrants armed with *kieries* (clubs with knobbed heads).[26]

This event led to the creation of a 'black reserve guard' within the South African police force, whose members wore the uniform and carried the weapons of the South African police and performed the same tasks, but as unpaid volunteers.[27] The reserve force had been set up in 1961 to help the South African police enforce the countrywide state of emergency after the ANC and the PAC were banned: it took over some policing functions while part of the forces were occupied with 'the fight against terrorism'. Although it was initially reserved for Whites, this 'grassroots' law enforcement was turned into an auxiliary police force in coloured areas, and later, beginning in 1966, for Africans, particularly to patrol the townships. Being a reservist had its advantages: one could obtain a driving license and learn to shoot, wearing a uniform and carrying a gun brought benefits in kind such as being 'offered' meals or beverages.[28] Within five years of the law's passage, the township of Guguletu had recruited 150 reservists.[29] Similarly, members of the Peacemakers group were integrated as reservists starting in 1976. Taking their orders from the police meant relinquishing the autonomous action of the vigilante associations, which, according to local leaders, used even more repressive methods than white civil servants in the police force.[30] In short, volunteering for a vigilante associations prepared many men for volunteering within the police, an intermediate scheme between the vigilante economy and the institutional police that was to play a preponderant role at the end of apartheid.

Violence and Vigilantism in South-West Nigeria

In Nigeria, the civil war (1967–1970) and the 1970s are often seen as the starting point of the history of violent criminality due to the supposed growth in the number of armed robberies that began during these years (Idowu 1980, p. 12; Inyang 1989, p. 75; Jemibewon 2001, p. 79; Tamuno 1989, pp. 92–93). Most of these studies underestimate the situation before the civil war, and take into account only a very short period during which criminality is always perceived to be increasing. It is common knowledge that police statistics reflect policing activity or fluctuations in the number of formal complaints more than a hypothetical rise in crime. In Nigeria, those statistics underestimated the number of armed robberies in certain periods (e.g. the late nineteenth and early twentieth centuries, for want of law enforcement), whereas their growth during the 1930s was partly linked to the recent creation of the Criminal Investigation Department that began recording crime. In any event, even a straightforward reading of the statistics indicates that one cannot trace their origin to the civil war: compared with the population increase, the number of armed robberies recorded by the Nigeria Police Force (NPF) doubled between the 1930s and the 1940s, and doubled again between the 1940s and the 1970s.[31] It seems more likely, then, that the consequences of the civil war merely exacerbated phenomena already observable for several decades in most Nigerian cities, notably the gradual replacement of knives or locally manufactured guns by automatic weapons; the application of the death penalty in cases of armed robbery starting in 1971, which encouraged robbers to kill victims and witnesses; and the spread of armed robbery to the newly prosperous areas of the oil economy (Chapter 4). Beyond these specific circumstances, the feeling of not being protected is part of a longer history rooted in the uneven nature of policing and the emergence of relationship between the population and the police.

The duality of Nigeria's police system imposed *de facto* limits on its operationality across large swaths of territory. Under indirect rule, police working under native authorities worked alongside more centralised police forces. The merger of the colonial police of Lagos and the Northern and Southern provinces culminated in the creation of the NPF in 1930, paid and trained by – and under the orders of – the British administration. The new police unit made up the principal police force in the new centres and urban spaces of the colonial state as well as in government residential areas (GRAs) and foreign and migrant neighbourhoods (the *sabo* and *sabon gari*). Its main activities – law enforcement, riot prevention and controlling the opposition – were carried out on a countrywide scale. Whenever possible, the British administration kept native police forces in rural areas and older urban centres, in line with their three-pronged agenda of keeping costs down, ensuring the loyalty of traditional leaders and building the native administration's capability for action (Owen 2012, pp. 25–51). The limited number of agents with little knowledge of modern police investigation methods were unable to prevent harm to persons and property in the territories under NPF authority.

Furthermore, there was tension within law enforcement institutions themselves between two opposing conceptions of order: in the 'orthodox' view, establishing professional police forces corresponded to protecting the public good; in the 'revisionist' or radical view, the purpose of the police was first to protect the interests and values of the ruling and property-owning class (Owen 2012, pp. 29–30).[32] Instead of seeing these two interpretations of the police as irreconcilable, Olly Owen suggests that the two conceptions coexisted – not without tension – within state power itself (Owen 2012). The repeated calls by residents to send police officers into sensitive areas reflected a need for protection far greater than any fear of coercion their presence might inspire within the population.

To understand the wider context of policing in the Southwest Region of Nigeria, three factors must be taken into account. First, the prevalence of crime and slave raids in the nineteenth century and their role in the history of recurring conflicts between city-states to gain control of trade routes in the region (Falola and Oguntomisin 1999). The Pax Britannica managed, though not without difficulty, to impose a ceasefire between rival city-states, and in many respects, the armed robberies carried out at the start of the twentieth century seemed to be a continuation of earlier conflicts under a different name, particularly in Ibadan, where a number of rival warlords lived (Watson 2000). Second, in the late nineteenth and early twentieth centuries, when Nigeria's economy had become more solidly integrated into the British colonial economy, new forms of poverty involving unemployment, proletarisation, prostitution and delinquency gradually supplanted its previous guises in the form of servitude, disability and famine (Illife 1987, p. 164). The cities of Lagos (32,000 inhabitants in 1891; 665,000 in 1960) and Ibadan (120,000 inhabitants in the late nineteenth century; 600,000 in 1960) pioneered this shift (Fourchard 2005, pp. 219–319; Illife 1987). Third, Nigeria entered into a long economic depression between 1914 and 1945 that was experienced more harshly in the South, which had become more integrated into the world economy, especially after 1930. The Nigerian press reported an increase in the number of petty thefts, armed robberies and burglaries in the areas with a high percentage of migrant residents on the outskirts of Lagos (Ikeja, Agege, Mushin) and in the pre-colonial historic centre of Ibadan (Fourchard 2005, pp. 219–319).

The boomtown of Agege (few inhabitants in 1914; 10,000 by 1935), at the northern periphery of Lagos in the heart of the cacao region, was emblematic of this radical transformation. The wages of day labourers on cacao farms dropped by two-thirds between 1927 and 1934, as cocoa planters were no longer able to sell their production after prices collapsed in 1929. District court records for the 1930s show a rise in thefts committed by peasants, craftsmen and tradesmen 'without means of support', and day-to-day violence consisting of multiple interpersonal clashes (tax collectors attacked by tradesmen, penniless passengers by hauliers, tradesmen by *jaguda* boys, etc.), a sign of social precariousness and the absence of local institutions capable of settling these conflicts.[33] A decisive factor in the emergence of a vigilante economy was the increase in night burglaries and a series of armed robberies between 1930 and 1934, prompting the local administrator to say that Agege harboured more criminals – past, present and yet to come – than any other city of comparable size in the world.[34] No doubt this was an exaggeration, as the burglaries were simple operations involving one or two people who pried open a door under the cover of night, stole things and ran away. Victims feared not only material loss in a time of crisis, but also the powers the burglars were held to possess: they were usually 'protected' by two types of charms, the first carrying the power not to awaken victims, and the second the power to paralyse them. Like nineteenth-century warlords, burglars (and their victims) believed in the efficacy of charms (Falola 1995, p. 10). The British administration apparently agreed and criminalised their use: in the Ikeja district alone, 40 people were found guilty of possessing these charms during the 1930s.[35]

It was in these urban territories that vigilant protection first arose; it was provided by hunter guards, a nineteenth-century force legitimised by colonial rules. The hunters, known in Yoruba *as ode*, were employed as night watchman by warlords in city-states during the nineteenth century. In the early twentieth century, they were eliminated by the British administration, which replaced them by the Native Authority (NA) Police (Falola 1989, p. 318; Watson 2003, pp. 76–78). The hunters resurfaced first in villages along the current border with the French colony of Dahomey (today Benin Republic) to combat smugglers during the 1920s, and later in the outskirts of Lagos during the 1930s to contain a wave of burglaries. The Second World War context noticeably heightened the need for protection using methods similar to those adopted when the civic guards were created in South Africa. Many lower-ranking African officers were enrolled *en masse* in the army, whereas the members of the NA Police were assigned to protect civilians during air raids (Owen 2012; Annual Report of the Nigeria Police Force 1940). Several spates of burglaries prior to and at the start of the war led to the recruitment of hunters equipped with charms to counteract burglars and armed robbers in the main cities of the region (Lagos, Ibadan, Oshogbo, Ede, Ogbomosho, Ife and Oyo).[36] In 1941–1942, the new system was regulated in the cities of Ibadan, Lagos and Oshogbo, taking the name of the earlier system from which its members were derived: the hunter guards. Their task was to ensure the enforcement of curfew under the authority of neighbourhood leaders and the city's native authorities.[37] Parallel to the hunter guards, voluntary associations known as *sode sode* were being created in certain neighbourhoods of Ibadan. In Yoruba, *sode* is the contracted form of the verb *se* (which means 'to act as') and its object *ode* (which therefore means both 'hunter' and 'guard'). *Sode sode* can thus be translated as 'act as the guard', 'act as the hunter' or 'constantly surveil'.

These organisations, originally envisaged as temporary solutions to offset the lack of police forces in wartime, were kept in place with more supervision after the war. In 1948, the Resident or colonial administrator in Ibadan, the headquarters of the Oyo province, laid down the rules authorising the creation of a group of hunter guards: a record of participants was to be kept; the hunter guards had to be supervised by a neighbourhood leader; firearms would be used only in self-defence; suspects would be handed over to the nearest police station; and no uniforms were to be worn.[38] Despite these constraints, demand for security was so strong that, within a few years, 78 groups of hunter guards were authorised to patrol in the Ibadan division alone, thus becoming one of the main forms of auxiliary police in the region.[39] The fact that these groups were operating in the historic centre of Ibadan shows that, in spite of attempts at supervision, their autonomy allowed them to regulate social life using coercion.

Honour and Violence in the Centre of Ibadan

Unlike the Cape Flats, which were products of the segregationist policy of apartheid, Ibadan's historic centre had been largely spared by the intervention of British engineers and planners. Located 120 km north of Lagos, the city was founded in 1829 by warlords engaged in a struggle against neighbouring city-states for control over trade routes between the coastal slave ports, Lagos and Ouidah (now in Benin) and the prosperous economy of the Caliphate of Sokoto (now Northern Nigeria). By the end of the nineteenth century, Ibadan had become one of the largest cities in Africa (its population was said to number 100,000) mainly soldiers, refugees, slaves and traders.

Following the British conquest, indirect rule was imposed in Ibadan as well as in the rest of colony. In this instance, the Native Authorities retained their power over policing, taxation and the justice system in the pre-colonial city-centres. The limited action of the colonial administration in Ibadan – its main accomplishment in 60 years was the construction of a road to open up the city's historic neighbourhoods – was careful not to disturb the social hierarchies in the areas dependent upon the Natives Authorities.

The Bere neighbourhood, where my investigation was situated, comprised large residential units (compounds known as *ile*) overseen by lineage chief (*mogaji*), who could be warlords (*balogun**) or city administrators (*baale**). In the nineteenth century, the fortunes and wealth of these chiefs were based on their valour as warriors, the number of their followers and family (wives, children, slaves, peasants and soldiers, which could include as many as 2,000 people), and on the prestige of their lineage praised in paeans known as *oriki*, sung in their honour (Barber 1991, p. 203). The sense of honour based on the public recognition and social esteem (*ola*) of the city's important families was a core ingredient in the moral economy of the historic neighbourhoods of Ibadan. In the nineteenth and early twentieth centuries, some of these chiefs preferred to commit suicide rather than undergo the dishonour of exile to preserve the honour of their lineage and their compound (*ola ile*) (Adeboye 2007). Under the *Pax Britannica*, the *bale* or neighbourhood chief acted as the neighbourhood relay of the Olubadan, the head chief of the city, who became the privileged interlocutor of the colonial administration. As the city assumed a larger role in the colonial economy and the property market expanded, the *ile* (or compounds) built in the nineteenth century were divided into numerous housing units for tenants who were not always dependents of the chiefs (Mabogunje 1962).

All the same, the chiefs retained a great deal of authority over all residents: they were the main organisers of the *sode sode*. These groups, contrary to those in Cape Town, were not set up to cope with the danger of supposedly unruly youth, but rather in response to a traumatic event, most often an armed robbery or night burglary. The *bale* asked the heads of households to designate volunteers to patrol the neighbourhood at night ringing a bell and calling out '*Ko n ilé gbelè*' ('Stay in your homes, the guards are outside').[40] The system was based on the residents' voluntary acceptance of the curfew and the ability of the heads of households to mobilise the youngest members and ensure their safety. The *sode sode* were equipped with clubs, whistles and above all charms provided by the *bale* to guarantee either protection (make people invisible and stop bullets), or far-reaching police powers (one could paralyse fleeing thieves or kill a suspect by putting a charm in his hand).[41] These were the rules of the neighbourhood and they were applied in the absence of the NA Police.[42]

Thus, police practices differed considerably from those at the Cape Flats: at the request of the *bale*, the *sode sode* patrols were made up mainly of the youngest men; they operated out every night of the week, controlling entries, exits and all movements within the neighbourhood. Once residents felt safe again, the patrols were abandoned, to be revived if necessary. Insecurity was never fundamentally perceived as coming from the inside. The area offered a relatively protected environment, due to the particular morphology of the historic centre (made up of narrow streets and lanes frequented by one's neighbours, with few public spaces or roads and no places to work, but a familiar neighbourhood) and the social control exercised over the residents by the *bale* (Figure 3.3). According to Akin L. Mabogunje, although these historic centres were described as slums because of the poverty of family households and the lack of basic services and sanitary living conditions, they were not areas of social deviance, criminality or delinquency in the early 1960s (Mabogunje 1968, p. 235). Unlike the Cape Flats where moral communities were created in opposition to delinquent youth, in Ibadan they were built on the authority of *bale* capable of protecting residents from perceived threats such as outsiders in the neighbourhoods (armed robbers, armed burglars or *jaguda* boys).

A sense of honour impelled the main lineage chiefs to resort to corporal punishment and exile to keep the children and youth of the neighbourhood under tight control: 'When we were young, if a child was a thief, once the people in the neighbourhood realised it, they informed the family, which was responsible for getting him back on track. If the child persisted, the family drove him from the neighbourhood and the city to preserve their reputation and honour'.[43] As for the others, namely those who came from elsewhere, there were varied and often more radical punishments: hours of compulsory agricultural work in local forests to benefit the *bale*;[44] mandatory expulsion of the offender to his or her village of origin; on-site execution using amulets or firearms, or

FIGURE 3.3 Bere, the historic centre of Ibadan (2003)
Source: Photos by Andrew Esiebo.

banishment to a neighbouring locality where there was a man in the habit of executing armed bandits.[45] Flagrant violations led to the use of summary justice for minor offences: the press often reported 'strollers' being beaten to death because they walked through neighbourhoods at the wrong time.

Once these organisations were granted an authorisation by the administration, they enjoyed almost total autonomy. This delegation of authority was not without risks. At the end of the 1950s and during the 1960s, they were involved, along with the NA Police, in the political conflicts of the Northwest Region, particularly the rivalry opposing the leader of the Action Group (AG), Obafemi Awolowo, to his regional enemy, Ladoke Akintola, a former AG member, Prime Minister of the Western Region starting in 1963, and founder of a new party, the Nigerian National Democratic Party (NNDP), an ally of the Northern People's Congress. The local authorities, the neighbourhood chiefs, the native police and the hunter guards became instruments of NNDP domination. The guards were used as an auxiliary force by the NNDP in its struggle against the AG before and during the 1965 elections, despite federal government injunctions to prohibit them.[46] They took part, along with the thugs recruited by the NNDP, in rigging the elections, setting off an unprecedented wave of political violence in the southwest that finally ended in the assassination of Akintola and the military coup d'état in January 1966 marking the end of the First Republic. Due to its involvement in the violence, the NA Police was eliminated and absorbed into the NPF, whereas the night guards were banned and their guns seized. As in South Africa, they survived the ban but fragmented, acting on a neighbourhood scale before once again being regulated by the federal state under the name of vigilantes in the late 1980s (Chapter 4).

Indeed, it was because the residential spaces for workers and migrants were peripheral to the smooth functioning of the colonial and industrial economy that employers shifted the responsibility for their well-being onto the government, and the government in turn authorised the populations to take charge of protecting them. Thus, vigilantism in these urban areas seems to have played a marginal role in the political economy of colonialism. The history of law enforcement in these outlying areas should therefore be analysed not by downplaying the importance of the economic stakes, but rather by trying to account for issues that were peripheral to the colonial economy.

Conclusion

These vigilante organisations could be seen as the manifestation of a shifting – though on-going – government technique: they negotiated the terms of their interventions with the administration, participated in a certain form of law enforcement and supplemented police forces busy with tasks judged more important at certain times. The technique remained indirect: security was provided exclusively by local organisations in Nigeria, and to a large extent in South Africa as well, although sometimes with the help of the police in townships. In Nigeria, it was based on residents' knowledge of the neighbourhoods, which were still viewed by the administration as relatively nebulous areas at the end of the colonial period. In South Africa, on the other hand, the technique was based on more precise data on the populations made possible by apartheid stabilisation and relocation policies from 1950 to 1970, and by monitoring social disorder, crimes and offences, which were regularly reported, thanks to information provided by vigilante associations. As such, they were all involved in 'doing the state', processes which could occur in cooperation or competition with stage agencies (Migdal and Slichte 2005, pp. 14–15, quoted in Hoffmann and Kirk 2013, p. 11).

In both South Africa and Nigeria, the state authorised these organisations to resort to violence as long as it did not go against local, national or imperial interests. Offences against persons and property – recurrent in South African townships and occasional in Nigeria – combined with an under-developed criminal police, led the administration to delegate or discharge substantial powers to these groups: an authorisation to use local firearms in Nigeria, an authorisation to use corporal punishment in South Africa. The colonial administration, and later the Nigerian state, tolerated the violence practised by these volunteer groups provided they abided by the authority of the neighbourhood chiefs or local associations. As soon as they appeared to be the armed wing of a political party threatening state security, they were banned and disarmed (the Civic Guards and Peacemakers in South Africa, the night guards in Nigeria). These organisations were less an example of local sovereignty than a form a public authority sanctioned by the colonial administration, which passed its police functions on to various local groups.

A recognition of the twilight nature of vigilante organisations implies to investigate who is included and who is excluded and what this implies for their access to justice and security (Hoffmann and Kirk 2013). Groups took over the operations of colonial and apartheid administrations, which their leaders claimed to be carrying out reinforcing their legitimacy and helping to legitimise colonial rule. In this co-existence of multiple public authorities within the colonial world, each instance gave its own meaning to authority. In delegating such power, the administrative authorities gave a set of organisations the chance to redefine the borders of their communities through violence if need be. The government's responsibility fell upon the shoulders of the neighbourhood leaders. Their actions were quite specific because they were designed to achieve their own local objectives: in seeking to strengthen neighbourhood cohesiveness, the organisations came to engage in violence against those seen as threats – migrants living in hostels, children and youths in South African townships, and outsiders unfamiliar with Nigerian neighbourhoods and their rules. These actions reinforced the authority and legitimacy of patriarchal organisations and the divisions produced by colonial and apartheid engineering (migrants vs. urbans, natives vs. non-natives, elders vs. young people). The increasingly commonplace recourse to violence as a method of social regulation on a neighbourhood scale seems to have echoed the violence shown against youth (and migrants in South Africa) inside the institutions of the coercive welfare state, which continually consolidated its hold after the Second World War (Chapters 1 and 2). In this economy, residents acquiesced to the acts of violence carried out by vigilante organisations because they ensured the protection and cohesion of a community that defined itself precisely in opposition to those threats. If there are few first-hand accounts of violence from victims in Nigeria or South Africa, it is perhaps because young people did not basically object to punishment-based supervision, which thereby guaranteed the continuation of these forms of coercive protection. But how could a protection economy, founded upon the relationship between generations, survive when that relationship to authority was called into question by youth who were either encouraged to volunteer as guards or brutalised by vigilante groups?

Notes

1. An expression borrowed from Thomas Sauvadet (2006). With regard to the problems of describing Cape Town *skollies* as gangsters (Jensen 2008, p. 98).
2. On the origin and contemporary evolution of this service within the South African state (Cabane 2012).
3. NASA, BAO 520/400 (12) 1942, Paul R. Mosaka, Member Native Representative Council Transval, Orlando Township to the Secretary for Native Affaires, Pretoria, 15 December 1942.
4. NASA, BAO 520/400 (12) 1950, Native Commissioner to Director of Native Labour, 14 December 1950, Harry L. Mekela to the director of the Rand Daily Mail, September 1951.
5. NASA, BAO 520/400 (12), Chief Native Commissioner to the Secretary for Native Affairs, 8 March 1951; Henri Philippe Junod, 'Labour Bureaus and "Vigilantes" in African Townships', *Penal Reform News*, 20 January 1952, pp. 8–10.

6. Civic guards could be found in the Rand townships of Stirtonville (1956), Chiawelo (1950s–1960s), Zola (1960–1965), Orlando (1965), Mofolo (1965–1967), Natal-spruit (1955 and 1969), Springs (1969) and Tladi/Moletsane (1969–1972). (Glaser 2005).

7. 'Violence Survey Social Chaos's on the Cape Flats', *Cape Times*, 1 May 1970.

8. *Living in Manenberg and Bonteheuwel. Workshop on the living conditions in some Coloured Townships on the Cape Flats, and their implications for the incidence of deviant behavior*, Cape Town, National Institute for Crime Prevention and the Rehabilitation of Offenders.

9. Interview with Aefan Davids, a former Peacemaker, Manenberg, Cape Town, February 2009.

10. Interview with Robin Roberts, a former Peacemaker, Manenberg, Cape Town, May 2009.

11. ACT, 3CT 1/4/18/1, Suburban management 1974–1976. Athlone and District Management committee minutes, 1974–1976.

12. The National Institute for Crime Prevention and the Reintegration of Offenders (NICRO) is an influential national association financed by the Ministry of Justice.

13. 'We talked to the youth, but they didn't want to understand that we had to punish them. Whatever else we said to them, they were told not to do it again or they would get a good thrashing. Community members must respect their elders; respect their parents and authority', Interview with Aefan Davids, former Peacemaker, Manenberg, Cape Town, February 2009.

14. This was the case of the following association: Ilizo Lomzi Nyanga West, Vigilant League of Decency in Guguletu, Coordinating Committee to Combat Juvenile Delinquency of Langa, Guguletu Civic and Welfare Association, and Community Services Association of Guguletu.

15. AWC 2/6, Nyanga West Vigilance Association to Native Administration Office, Langa, 5 September 1961.

16. 'Nyanga Move against Loiterers', *Cape Times*, 7 November 1961.

17. 'Let us Form Civil Guard, Say Nyanga Residents', *Cape Times*, 25 September 1961.

18. 'Nyanga Move against Loiterers', art. cité.

19. AWC 2/8, Constitution of the Vigilant League of Decency of Guguletu, 21 September 1964.

20. *Ibid.*

21. AWC 2/8, Director Bantu Administration to the Secreatry Ilizo Lomzi, Guguletu Section, 19 August 1965.

22. Interview with Abie Sumbulu, Bordeaux, March 2015.

23. Interview with Abie Sumbulu, Bordeaux, March 2015.

24. A similar merging process took place in numerous other situations in Africa and across the world (Abrahams 2007, p. 426).

25. Interview with Tabile Ndabeni; March 2009, Cape Town.

26. 'Labour Compound Urged for Langa Peace', *Cape Times*, 27 June 1966; 'Murders in Langa set off Fighting', *Cape Times*, 27 June 1966.

27. AWC 3/52, Secretariat Bantu administration in Langa to city of Cape Town, 12 July 1966; 'Langu Guards Ready Soon', *Cape Times*, 5 July 1966.

28. Interview with a women reservist in Manenberg, June 2009.

29. AWC, 3/52, Director of Bantu Administration to the town clerk, Guguletu, 12 October 1971.

30. AWC, 3 CT 1/4/18/1/1/5, Crime in the management committee area, 23 November 1978.

31. Ten cases were reported per year during the 1920s, more than 100 cases between 1936 and 1939, more than 200 cases between 1945 and 1947, and between 800 and 1,500 in the 1970s.

32. Tekena Tamuno (1970) upheld the first conception, Philip Ahire the second (1991).

33. NAI, Comcol 1, Ikeja District, Criminal Record Book, 1930–1940.

34. NAI, Survey Department, C. 69, Intelligence Report of the Ikeja District, 1935.

35. NAI, Comcol 1, Ikeja District, Criminal Record Book, 1930–1940.

36. 'In the Interest of Public Safety', *Nigerian Daily Times*, 3 January 1940; 'Robbers in Ibadan', *Yoruba News*, 9 April 1940; 'Hunters in Town', *Yoruba News*, 30 April 1940; 'Robbery in Ibadan and District', *Nigerian Daily Times*, 16 May 1941.

37. NAI, Comcol 1, 2498, Assistant Superintendent of Police to Superintendent, Nigeria Police, 1 August 1941; NAI, Ibadandiv, 1/1, 167, Hunter guard instructions, 5 January 1942.

38. NAI, Ibadandiv, 1717, Acting resident Oyo Province to District Officer Ibadan, 23 October 1948.

39. Similar organisations were operating in the main cities of the region, Lagos, Benin City and Oshogbo.

40. Interview with Wahabi Lawal, Ibadan, 14 January 2003.

41. Interview with Raliatu Adekanbi, 21 January 2003, Rasheed Aderinto, 22 January 2003, and Amusa Adedupo, 29 January 2003, Ibadan.

42. Interviews with Asiru Oluokun, 6 February 2003 and Amusa Adedupo, 2 January 2003, Ibadan.

43. Interviews with Alhadi G. O. Alatise, Ibadan, 23 January 2003.

44. Interview with Rasheed Aderinto, Ibadan, 22 January 2003.

45. Interview with Amusa Adedupo, Ibadan, 29 January 2003.

46. 'Night-Guards. Law Again?', *Nigerian Tribune*, 17 July 1965; 'Police Urged to Rebut Thuggery', *Nigerian Tribune*, 2 September 1965; 'Police, Council Clash over Night Guards' services', *West African Pilot*, 1 September 1965.

CHAPTER 4

Commodification, Politicisation and Uneven Pacification of Contemporary Vigilantism

Over the past 20 years, two crucial issues have perceptibly altered the terms of the security debate in both Nigeria and South Africa: the effects of political regime change – from apartheid to a democratic government in South Africa in 1994 and from a military to a civilian government in Nigeria in 1998–1999 – and the effects of neoliberal policies on the security arrangements of urban areas. Many interpretations emphasise the transformation brought about by these regime changes in relation to the efforts of the police to ensure popular consent. There are also considerable debates concerning the relative influence of national contingencies vs. neoliberal imported models on security sector reform. Together, this research opens up questions concerning the politicisation and state regulation of the older vigilante and communal security organisations as well as the emergence of new forms of 'privatisation' of security.

Conjunctural reasons have been put forward in both countries to account for the persistently high level of criminal violence after apartheid and since the end of military dictatorships in Nigeria. In South Africa, vigilantism remains frequently practiced despite a celebrated transition to democracy, a lauded constitution, and massive transformations of the state's legal apparatus following democratisation (Smith 2019). The violent socialisation in South Africa during the conflict of the 1980s and 1990s has been the usual reason given for the continuing 'culture of violence' after 1994, explaining the very high rate of homicides and vigilante violence in the post-apartheid period.[1] However, explanations suggesting the transformation of the so-called 'political' violence into 'social and criminal' violence in the post-apartheid period underestimates the impact of the routine use of violence prior to the conflict and fails to account for the continuation of those practices long after the conflict ended (Kynoch 2005a, pp. 493–494).

Classify, Exclude, Police: Urban Lives in South Africa and Nigeria, English Language First Edition. Laurent Fourchard.
© 2021 John Wiley & Sons Ltd. Published 2021 by John Wiley & Sons Ltd.

In Nigeria, the proliferation of militias and vigilantes is often linked to the restoration of a civilian government in 1999 and is said to attest to an erosion of state authority and sovereignty (Bach 2006; Eberlein 2006; Young 2004). This reading of events falls into the trap of mistaking new organisations (new vigilante groups) for new phenomena (vigilantism as a longer historical phenomenon). It considers these often-violent organisations as the manifestation of the incapacity of the state to monopolise violence over its territory. This fails to take into consideration the delegation of policing over a longer period of time and uneven use of violence by vigilante groups in different parts of the country.

Research on neoliberal urbanisation in the African context shows, on its side, the development of a new urban order marked by the 'withdrawal' of the state and the simultaneous introduction of new strategies of social control, police activity and surveillance. Urban renewal policies are analysed as a means of transforming spaces of economic competition into controlled, sterilised and 'purified' spaces of social reproduction (Brenner and Theodore 2002, pp. 371–372). This 'splintering urbanism', made up of urban enclaves, infrastructures and fragmented services, leads to greater fear of the social mix on the part of the most privileged: 'The high demand for Business Improvement Districts (BIDs), pedestrian areas, shopping centres and leisure complexes has been met by purging urban space of poor people, crime and squalor'(Graham and Marvin 2001, pp. 215, 383). In South Africa in particular, this is visible in more aggressive policing of the poor and marginalised individuals (Samara 2010), the creation of fortified enclaves and the increasing polarisation of social groups (Lemanski 2004) in which the proposal of security services is only part of a wider project in which basic infrastructures, amenities and the regulation of common spaces are in the hands of private companies (Webert and Murray 2015). The development of City Improvement Districts (CIDs) in the CBD of Johannesburg and Cape Town in 2000–2002 are viewed as models of infra-municipal governance and autonomous tax zones that enhance the entrepreneurial appeal of central locations and secure properties, while excluding or relegating the poor and marginalised to the periphery (Didier, Peyroux and Morange 2012; Ward 2006). Finally, the growing adoption of zero tolerance policing policies is also crucial to urban regeneration programmes in Cape Town as in the historic neighbourhoods of Mexico City, New York or Cape Town (Abrahamsen and Williams 2009; Belina and Helms 2003; Müller 2016). This reading accounts above all for the most visible changes in business centres, shopping centres and urban enclaves even if for some scholars, this neoliberal urbanisation also concerns vigilantism and poorer areas (Comaroff and Comaroff 2006; Goldstein 2005).

While I agree that 'urban neoliberal order' is marked by more systematic surveillance, the development of urban enclaves, and delegation of security to citizens, it may miss other dimensions of urban policing in a low-income urban space. Private security companies are rare in low-income neighbourhoods and if citizens are increasingly involved in community police programmes seen as another form of neoliberal forms of policing (Brogden and Nijhar 2013; Garland 2001) – in Nigeria and South Africa, these often take over ancient forms of security mobilisations by local residents. In other words, the longer colonial and postcolonial trajectories of vigilante groups indicate the irreducibility of vigilantism to neoliberalism and weak-state analysis and instead open up a dialogue between urban studies and the anthropology and sociology of policing on cheap forms of urban policing in today's low-income neighbourhoods.

First, we need to take another look at the changes and continuities in urban vigilante practices since the end of the colonial and apartheid period. As suggested in Chapter 3, vigilante groups are twilight institutions. It is impossible to anticipate a single trajectory for their relationship to the state: the groups are not only a challenge to its authority; they may be absorbed by the administrative machine and their members turned into civil servants, or dissolved and banned by the police; they may operate autonomously but monitored from a distance, or their bureaucratic capabilities may be strengthened by the state through targeted financing. Focusing on the daily work of policing carried out by organisations today help to follow up questions asked in the former chapter. From the standpoint of the actors, the terms 'vigilante' and 'vigilantism' may mean different things with divergent political consequences. Tracing the semantic shifts stemming from transformations in the legal landscape, the criminalisation of certain practices, the ability of the police to monopolise violence in a given territory, and the historicity and memory of those practices embedded in their social and political use will give insights into the changing politics of urban security in our case study areas. Comparing the acceptance vs. rejection of the terms in South Africa and Nigeria shows how its evolving usage illuminates contested practices but it also assesses the transformation of policing on a nationwide scale. In South Africa vigilantism was banned in 1994 while in Nigeria vigilante groups were promoted in 1987 on the condition that they will not resort to violence anymore. Despite this new legal framework, the literature has emphasised the continuing use of violence by these groups in *both* countries. This assertion is open to question though and calls for a closer look at how state prohibition of violence affected concrete urban vigilante practices. Here we draw on interviews in the historic centre of Ibadan and interviews and ethnographic observations including participant observation in seven night-patrols with neighbourhood watch (NW) organisations in 2009 and 2010 in the coloured neighbourhoods of Cape Flats. Based on these evidences, I first suggest that the interdiction of uses of violence has changed the daily routine of vigilante organisations: this pacification is limited as numerous urban vigilante groups have stronger restrictions to use violence, which might however be used in specific circumstances or contexts. The chapter explores other past routines still embedded in today's daily policing such as the exclusion and classification of groups considered as dangerous for neighbourhood's major security organisations.

Second, the police in South Africa and Nigeria retain powerful symbolic, coercive and bureaucratic power (Owen and Cooper-Knock 2015). They have kept a central albeit contested role, in wealthy neighbourhoods and those in the city centre, often protected by private companies which can simultaneously undermine and support state agencies by their practices (Diphoorn 2016). CID partnerships have developed into something more nuanced than their original neoliberal strategy 'where the state and non-state have intermingling authorities and functions' (Berg and Shearing 2015, p. 91). Private security companies may hold the suspect until the police arrive; they may engage in joint operation with the police, share radio communication and fill paper work for the South African Police Service, a relationship which reveals the coexistence of multiple sites of authorities (Berg and Shearing 2015).

In low-income neighbourhoods, where private companies play a more marginal role, it is understandable that major police reform represented by community policing could be interpreted as the determination on the part of the police to increase their hold over the population and claim state sovereignty by domesticating non-regulated forms of law enforcement (Kirsch 2010; Marks, Shearing and Wood 2009, p. 146; Ruteere and Pommerolle 2003). Handing over the suspect to the police, patrolling with or on behalf of the police, doing its paper work and organising joint meetings with them are now common in the low-income neighbourhoods of Cape Town and Ibadan. What is less clear, though, is the ambivalent effects of the reform, which drew on embedded existing vigilante practices even as it transformed them. Beyond the issue of violence, there is a process of bureaucratisation and politicisation of vigilante groups at play. Bureaucratisation of security is part of a broader process characterised by the rise of technical rationality, the invasion of market and business standards, and the formal adoption of a form of remote government (Hibou 2013, p. 18). This bureaucratisation is not de-politicised. As the demand for security is very high, it remains an important political and electoral issue. Attempts at regulation in both countries, then cannot be grasped without analysing the potential politicisation of the security issue allowed by political parties. This politicisation does not explain in itself the resurgence of vigilantism in post-apartheid South Africa (Smith 2019); however, through the support of different schemes, political parties contribute to reshape vigilante interventions in townships. This process of politicisation is unevenly developed in the urban areas of both countries – more pronounced in the Cape Flats than in the centre of Ibadan – for a set of reasons that need to be explored in detail.

Finally, while vigilantism is acknowledged to be a cheap form of law enforcement (Pratten and Sen 2007), how it is financed has remained relatively obscure. The very expression 'privatisation of security' does not mean much with regard to low-income neighbourhoods where security companies are reluctant to invest or operate and where the state delegated security to the local population long ago. Rather we suggest that the process of *commodification* has been under way, through which a service once provided on a volunteer basis has become a paid service performed by various local suppliers and with some scope for state oversight. A shift has taken place according to a time frame and forms specific to each city. Increased financing of security in low-income neighbourhoods has in turn affected the kind of actors taking part in it, especially women marginalised in the labour market.

State Regulation and Commodification in Nigeria

In January 2009, in Ilorin, the capital of the state of Kwara, located 150 km to the north of Ibadan, a group of vigilantes stopped a goat and handed it over at the neighbourhood police station: according to them, the animal is an armed robber who had the power to turn into a goat just before he was arrested. A police spokesperson for the state of Kwara announced that the suspect remained in police custody until the end of the investigation. The police were reported to have paraded the goat in front of several journalists and some police officers tried to make it confess, forcing the spokesperson for the federal police, Emmanuel Ojukwu, to issue a necessary clarification two days later: 'Goats cannot commit crimes'.[2]

The story, that appears to be absurd, touches on two main issues raised by the forms of contemporary vigilantism in Nigeria. First, it evokes the popular imaginary shared by armed robbers, vigilante groups and the police, who are convinced that robbers possess supernatural powers that are difficult to counteract – a point already mentioned in the previous chapter. It is also noteworthy that the suspect who turned into a goat instead of being executed on the spot or beaten up by the vigilante group was voluntarily handed over to the police indicating that both the military and civilian regimes have been trying to regulating these organisations and bringing them within the purview of state institutions.

Following the Biafran War (1967–1970), the military regimes held on to power for 30 years (1970–1998), interrupted by a brief period of civilian rule (the Second Republic of 1978–1983). These regimes, initially supported by the oil boom and exponential growth of the state budget, were subsequently faced with a long period of economic depression (the 1980s and 1990s). A structural adjustment policy was imposed in 1986 and the state budget drastically reduced, as unemployment and poverty rose significantly.[3] While certainly relevant, this alone does not account for the transformation of criminal practices or the security measures taken by the federal state, which had a different rationale.

The oil boom brought with it a rise in the number of motorised vehicles driven by the population as well as by robbers, who changed their targets: holdups of automobiles on highways and in the streets became common occurrences in the 1970s, leading to the multiplication of roadblocks by the army and the police. Armed robbers became the embodiment of evil in the popular imagination precisely in the immediate aftermath of the civil war (Owen 2012, p. 42). Belief in the supernatural power of robbers led the military regime to hold their executions in public starting in 1971 to prove their 'magical protection' did not work. Nevertheless, the dissemination of their exploits in the press helped to popularise their relative invincibility; there seemed to no longer be any territorial limits on their areas of operation. Thanks to their newfound mobility, the robbers could target business districts, residential neighbourhoods, industrial and port zones, and even the coastline, where the first acts of piracy were recorded between 1975 and 1977 (Amirell 2009; Fourchard 2006b; Obbi 1989). By becoming de-territorialised, these attacks increased the feeling of insecurity among the chief beneficiaries of the oil boom: owners' associations were set up in residential neighbourhoods where they recruited private security companies and a new security-oriented architecture developed with a range of offerings depending on the homeowner's wealth and the financial strength and mandate of neighbourhood associations (Agbola 1997; Bénit-Gbaffou, Owuor and Fabiyi 2011). These residents' concerns became a political issue for the military governments whose legitimacy depended to a great extent on their ability to ensure order and security, the key to attracting foreign investors and preserving oil revenues.

Unlike the hunter-guards in the 1930s–1950s, the expansion of vigilante groups beginning in 1987 was not a response of elders or local authorities to the perception of localised insecurity, or even to the spread of the new forms of violence, but rather an action taken by the federal government to bolster the image of the police. In particular, the introduction of the vigilantes was the direct consequence of the media attention given to the saga of Lawrence Anini, the most famous armed robber in Nigeria. Throughout the year 1986, he and his men challenged the federal police in various locations in the country. The robber was presented by the national press as a social bandit a sort of Robin Hood figure and a symbol of the country's economic hardship, a narrative that the government of Ibrahim Babangida (1985–1992) took as a concerted attack on the regime's legitimacy (Marenin 1987). Anini's arrest and trial (1986–1987) revealed that he had enjoyed the complicity of the deputy chief of police of a Southern State, which confirmed in the eyes of the public and the press that the police was not only corrupt but organising the milieu of armed robbers.

To improve relations between the police and the public, the Inspector General of Police decided to grant communities the authority to form vigilante groups, while at the same time asking the media to stop disseminating sensational stories that alarmed the population and demoralised the police. Military governors were put in charge of setting up vigilante groups, which was initially carried out in the Oyo State in 1987. Unlike night guards who were banned in 1966, the vigilantes were protected by law: 'it is a group of persons acting at the neighbourhood or local government level to receive within its jurisdiction information on suspected criminals intended for the police in its struggle to detect and prevent crimes'.[4] Their duties were defined by decree and they were granted police powers: 'keep watch over residents' movements in the neighbourhood, maintain a registry of the guardians, respond to the distress of neighbours, arrest and hand over suspects to law enforcement officers, register with the police any residents who act as guardians and paid guards'.[5] The governor suggested that vigilantes also assume duties of social and administrative control: 'register tenants and property owners, examine the residents' business activities, monitor the movements of foreigners and individuals who suddenly become rich'.[6] The aim of introducing vigilantes in Nigeria was to restore the tarnished image of the federal police and regain control over the organisations that had survived the outlawing of night guards. We turn now to look at these developments in more detail in Ibadan.

Commodifying Protection and Regulating Vigilante Violence in Ibadan

To a large extent the initiative on the part of the Nigerian state to strengthen security through legitimising communally based vigilante groups had not basically changed the relationship between the police and the population (Fourchard 2008). Instead, it seemed merely to legalise and institutionalise older forms of security groups (especially the *sode sode* and the night guards). However, the introduction of vigilantes led to renewed cooperation between the police and the population, especially to regulate the forms of violence exercised by the organisations during the previous decades. The elders wel-comed this as a significant change, as residents could no longer take the law into their own hands:

> Around thirty years ago, the police were involved in training vigilante groups by teaching them how to arrest thieves.[7] When we had *ode*, there was no connection to the police. Now it's different: vigilante members are registered and officially recognised by the police to keep bad elements from joining their ranks. The aim of the police was to curb excesses (e.g. killing enemies, the armed robbers).[8] When we started the vigilantes, we consulted with the police. They checked what we were doing from time to time, and anyone we arrested had to be handed over to the police, and then we told them what had happened.[9] Today it is impossible to kill a thief. Even if you beat a thief, the police could harass you for it; what we do now is turn thieves over to the police.[10] Today the *ode* are better prepared than in the past. We have advised *ode* not to shoot at thieves or wound them, but rather to try and arrest them so they can be handed over to the police.[11]

Aside from these testimonials, it is hard to ascertain whether earlier, more violent forms of justice actually declined or not; additional investigations are necessary here. Nevertheless, all those interviewed testify to a change in rela-tions between the police and the population. Henceforth, neighbourhood leaders must submit a file including the photos and identity cards of vigi-lante members to the police station for validation and permission to oper-ate, a procedure reminiscent of registering hunter-guards with the colonial administrator during the 1940s. In contrast to the colonial period, however, today's neighbourhood representatives hold monthly meetings at the police station to exchange information; the crimes and offences committed in the neighbourhood are itemised and the police present an assessment of crimi-nality in their jurisdiction. It appears that residents must pay to benefit from these services (see below).[12] Finally, the introduction of vigilantes has resulted in a more obvious police presence in the neighbourhood. Regular patrols are conducted in exchange for payment made by neighbourhood leaders to the police station, attesting to the fact that, as a matter of priority, the police pro-vide their services to those who can afford to pay. A 2002 study indicates that police controls are generally effective even though they may not be frequent (Fourchard 2003).

These new regulations have not changed what residents identify as threat-ening to their security. Danger is still associated with outsiders, whereas the young people within the neighbourhood are seen as non-violent and low risk:

Those who steal at night come from outside, whereas those who steal clothes or animals during the day are from the neighbourhood.[13] The danger comes from areas boys, but they are not considered as part of our community.[14] We cannot say that these hooligans are from Bere: some of them come from Olomi, Olodi, and Aparta; they are transported to Bere and at night they go back home. If they came from the neighbourhood, we would ask their parents to take them to task.[15]

The main differences in relation to the earlier period come from the fact that vigilantes can no longer independently decide to physically eliminate or punish the individuals that embody this menace, and that paid guards have replaced volunteers. The transition from a volunteer-based protection model to a remuneration-based model is all the more remarkable in that there is practically no wage-based economic activity in the historic centre of Ibadan, which has about one million inhabitants today.[16] Street trade and markets account for 70% of the economic activity of the metropolis and even more within the historic centre, where there are very few wage earners and, unlike the colonial period, almost no peasants making a living from cacao cultivation (Fourchard 2003, p. 5). The level of poverty is extremely high: at the time of the 2000 survey, 65% of households earn less than 35 US dollars per month (far below the minimum subsistence wage set by the federal government at 50 US dollars per month in 2000) (Fourchard 2003). But weak monetary circulation combined with a relatively reduced feeling of insecurity has not prevented the commodification of protection.

The *sode sode* were based on the elders' ability to mobilise youths and to offer them efficient protections against armed robbers. Those patrols were abandoned at an unknown date, probably around the 1970s or 1980s, and replaced by paid guardians ('*ode*' or 'vigilantes', the terms used to designate this activity since 1987). Night guards and *sode sode* were allowed to carry local weapons, but the vigilantes introduced by the police are no longer allowed to do so, which has considerably upset the balance of power between aggressors and protection groups. Together, the development of more violent criminal practices and the circulation of automatic weapons since the end of the civil war diminished the power of the charms provided by the *bale*, the local chief. In this part of Nigeria, as in Cameroon, the magic powers of these elders have not been employed to resist change (Geschiere 1995). It would be more accurate to say that certain specialists in making charms were able to adapt to more violent practices by making more efficacious charms (see below). Other, more ancient charms are still in use today: 'We have village charms that are attached to the front and back of our homes to ward off theft. Now if a thief comes he cannot leave with what he has stolen'.[17] This object referred to as *àálé* is commonly used in the region to protect properties and serve as warnings for thieves (Doris 2011). On the other hand, the elders agree on the difficulty or inability of the charms supplied by the *bale* to adjust to modern weapons. 'The guns made locally had no effect on the *sode sode*. The charms could control them, but that is no longer possible with foreign guns: they kill!'[18] Thus, many elders regret that a number of charms no longer work or that neighbourhood elders who once knew how to prevent attacks have since died. They also lament that there are so few volunteers or that they pretend to be ill. 'Due to the lack of communitarianism, the *sode sode* are no longer effective. Before, people respected the opinion of the *bale*, but they don't anymore'.[19] This loss of public spirit attests to the weakening of the lineage chiefs' authority to guide the conduct of young people, especially to perform time-consuming, exhausting, and dangerous tasks of the *sode sode*. It also reflects the long-term process of individualisation that marked twentieth century urban Africa, and Ibadan in particular, through the development of the cacao economy, conversions to Christianity or Islam, and the subdivision of compounds into individual houses occupied by smaller families (Mabogunje 1968). In short, doubt about the ability of charms to provide effective security to group members has led to questioning the authority of the neighbourhood and lineage chiefs, whereas the replacement of the *sode sode* by paid guardians shows the relative loss of social control on the scale of the immediate vicinity or the whole neighbourhood.

Henceforth, effective protection requires the financial participation of every household. Those accustomed to patrolling at night when they were young now make monthly payments.[20] A form of localised citizenship has thus emerged, based not on the voluntary service provided but on a form of individual tax, and the ability of the head of the family to pay it. Each family unit contributes between 70 and 100 nairas a month (50–70 euro cents) towards neighbourhood guards. The centre of Ibadan is one of those urban areas in which lack of money (translated by the popular expression '*kà s'owo n'igboro*', i.e. 'there is no money in town') is the main concern (Guyer and Salami 2011). In this context of penury, the money supply is ensured on a very small scale through deferred payments and tontine practices are used to fund most everyday activities, including protection in this case. These exchanges imply relationships of trust and reciprocity and new forms of commitment in which the ability to pay the amount due, no matter how small, is necessary to preserve family honour: 'If a family refuses to pay, we announce the name via the public address system at the mosque. Once this happens, the family will soon make its payment'.[21]

Neighbourhood guards receive between 4,000 and 6,000 nairas a month (30–40 euros). The low wages reflect the income level of neighbourhood households. The sizeable difference in the wages paid to each *ode* depends on the number of compounds (and on the number of houses inside the compound) to be watched during the night. For example, an *ode* is paid 5,000 nairas to keep watch over 27 compounds in the neighbourhood.[22] The moral community that had emerged during the colonial period to provide security for the entire neighbourhood has been replaced by the individualisation and commodification of the protection in which the ability to pay now substitutes for the ability to supply volunteers. No doubt that is why one elder suggested: 'We began to pay the *ode* when modernity arrived'.[23] These two major transformations – the commodification of vigilantism and greater police regulation – have not been limited to the micro-social scale; they can now be found on a metropolitan, regional and national scale, particularly since the return to civil government in 1999.

Return to Democracy and Uneven Pacification of Vigilantism

Numerous studies have asserted that the return to democracy was accompanied by an increase in militias and vigilantes. Independently of the Oodua People's Congress (OPC), the so-called 'ethnic' militia created in 1994 to protect Yoruba interests in the South-West, armed groups in the Niger delta began multiplying in 1998, vigilante group like the Bakassi Boys were set up the same year in several Eastern cities (Aba, Onitsha), and the Islamic police (*hisba*) have been introduced in most of the Northern states since 2000. Various interpretations have been put forward to account for this chronological conjunction. Two explanations predominate in the analyses: the first argues that the creation of vigilante groups was a response to the increase in criminality during the 1990s and the return to democracy in 1999, which lessened military presence in cities (Agbu 2004; Harnischfeger 2003, p. 27). The second maintains that the ethnic militias were created by civil society groups that had been struggling to establish democracy since the mid-1980s and then yielded to the temptation of 'ethnic nationalism' once democracy had returned (Aiyede 2003; Ikelegbe 2001). These readings adopt a mildly teleological slant (i.e. viewing the creation of the vigilantes as the culmination of necessarily rising criminality) and present a short-term view that sees vigilante organisations as a new phenomenon. There are, however, two additional phenomena that add complexity to this overly linear history: as we have evidenced now in great detail, the militias and vigilantes are often linked to earlier organisations; they are in some instances more regulated by a state capable of turning certain organisations into semi-public institutions if need be.

Thus, the new organisations in the 1990s do not mean the phenomenon is new in Nigeria. In Kano, the vigilantes of the 1980s replaced the Yantauri organisation, which had been active for several decades (Olaniyi 2005b, p. 25). An organisation similar to the Bakassi Boys operated at markets in Eastern cities during the 1970s and 1980s (Oneyonoru 2003) and the protection service provided by the OPC in the South-West since 2000 has many points in common with the one made up of vigilantes in the 1980s (see the following text). All these organisations have a limited lifespan and disappear when the feeling of insecurity wanes, when either human or financial resources become rare, when the state decides to dissolve them, or when they lose their legitimacy by becoming excessively politicised or criminalised. Sometimes they are combatted, but far more frequently they are tolerated or even promoted by state governors because, like their predecessors in the colonial era, they enable the authorities to pass on to the population the cost of their own security.

Since the elimination of regional police forces in 1966–1970, demand for local police has continually resurfaced, notably since 1998–1999, in states controlled by opposition parties or governors of rebel factions of the presidential party, which denounce 'the lack of federal police neutrality' and view the groups as supplementary forces to help achieve their electoral conquest. The Constitution of the Fourth Republic prohibits the formation of a police force within the states, but in fact many governors circumvent the law by setting up vigilante groups modelled on the organisation launched in 1987: the governor of the State of Osun created his own group of vigilantes in 1996, followed by the governors of the States of Niger (Niger People's Congress) in 2000, Edo (Edo State Vigilante Service) in 2001 and Adamawa (Neighbourhood Watch Organisation) (Sesay et al. 2003, p. 19). The governors of the states of Abia and Anambra turned the Bakassi Boys into paid state agents starting in 2000, before officially abolishing them in 2002 (Meagher 2007). In numerous other states, an official card delivered by the federal state is required to set up a vigilante group (Owen 2012, p. 47). The most significant case is no doubt the State of Plateau. The Vigilante Group of Nigeria (VGN), created jointly by the citizenry and the state in 1999, has been continually strengthened to take part in routine law enforcement and later to cope with the recurrent mass violence in this state since 2001 (Higazi 2016; Lar 2015). This co-production has been visible for 15 years in the examination of vigilante candidates by police officers, joint patrols, regular information meetings and similar uniforms and hierarchies. According to Jiman Lar (2015), the VGN are the result of a process of institutionalised vigilantism that began well before 1999.

Since 1987, the federal authorities have formally separated the ethnic militias they oppose from the vigilantes they are trying to regulate. In practice, a militia may be transformed into a vigilante group, as demonstrated by the OPC, the most powerful organisation is the South-West, which was created after the annulment of the 1993 election results that proclaimed victory for Moshood Abiola, a Yoruba candidate. Set up to fight against the political marginalisation of the Yoruba within the federation, the OPC was considered an ethnic militia at the time and was quick to show its opposition during the violent clashes between the federal police and the country's other ethnic associations. Thanks to the election of President Obasanjo in 1999, this militia was transformed into a group fighting criminality and corruption (Guichaoua 2009, p. 1660). At the time, many members offered their services in the poor neighbourhoods of South-Western metropolises. Indeed, the protection proposed by OPC members was almost identical to that of the *ode* and vigilantes in the historic centre of Ibadan: two to four members patrolled the neighbourhood at night and enforced curfew, for which they received a monthly salary of 5,000–8,000 nairas, usually paid by a contribution from the inhabitants of the neighbourhood (Akinyele 2007). The spiritual counsel given by the OPC to its members involved initiation ceremonies that were part of the long history of protection, which also fostered economic prosperity, good health and success (Guichaoua 2009, pp. 1660–1663; Nolte 2007). This additional physical, spiritual and moral protection and the reputation enjoyed by the OPC has transformed the organisation into one of the many security organisations in low- and middle-class-income neighbourhoods of Lagos (Revilla 2020). Finally, by changing its objectives, the OPC avoided the police persecution formerly directed against it. The organisation came to resemble a private security agency, providing protection for companies, pipelines in the Niger delta, parks and vehicles, public events and even the control of rubbish dumps in certain Lagos neighbourhoods by members looking to supplement their incomes (Salvaire 2019).

At first glance, the particular evolution of security services in the centre of Ibadan is consistent with the overall evolution of vigilantism in Nigeria, where the commodification of protection often goes hand in hand with greater government regulation. The expansion of the state budget between 2000 and 2015 made it possible to absorb more volunteers into the bureaucratic apparatus and increase the police presence in formerly neglected areas. This evolution has had its ups and downs, however, and the provision of security services to populations in poor neighbourhoods or in conflict zones is far from stabilised. In the end, the expression 'vigilante organisations' is an umbrella term that designates very different policing realities: anti-insurrectional groups that engage in massive amounts of violence in support of the army's struggle against Boko Haram in the Lake Chad region (ICG 2017); organisations that borrow police uniforms, hierarchies and bureaucratic codes, organise joint patrols with them, and provide them with regular reports, such as state vigilante services or the VGN found in states marked by recurrent conflicts like in Plateau state (Lar 2017). Or local groups that embrace the name 'vigilantes', which oversee social control and surveillance in a more pacified urban environment like in some popular low- and middle-class neighbourhoods of Lagos in more or less close cooperation with the police and in competition with other groups backed by the neighbourhood authorities of the local government or the federal government (Revilla 2020).[24] While 'vigilantism' refers to different realities in Nigeria, it does not have the pejorative connotation it does in South Africa.

Politicisation, Bureaucratisation and Feminisation of Vigilantism in the Cape Flats

From 1976 to 1994, various anti-apartheid movements violently opposed all those who were perceived as collaborators with the regime. This radical struggle politicised the former vigilante organisations but did not eliminate altogether the earlier models or the desire of some members to preserve a social order at the core of township life even in the face of the political storm. Starting in the 1980s, three types of organisations developed. The first and by far the largest on a countrywide scale, linked to the anti-apartheid movements (ANC, SACP, the Black Consciousness movement and the United Democratic Front – UDF), was made up of street committees, area committees and later People's Courts that drove the police from African townships like Alexandra: they did not set foot there again until several years after the end of apartheid (Seekings 1993; Steinberg 2008, pp. 49–57). The second included the radical wings of parties opposed to the ANC such as the Inkatha Freedom Party as well as armed groups called vigilantes supported and financed by the National Party to attack anti-apartheid activists (Cole 1987; Haysom 1986). It was precisely during the 1980s that the words 'vigilante' and 'vigilantism' lost their original positive connotation and acquired the pejorative meaning associated with anti-ANC movements who trampled on human rights. The third type was the Neigbourhood Watch (NW): although marginal on a countrywide scale, it dominated in the coloured neighbourhoods of Cape Town and was closest in purpose and activities to earlier movements. When apartheid ended, ushering in democracy and the promotion of human rights, vigilantism was outlawed: it had come to designate any act of violence committed by an organisation or individuals outside the legal framework approved by the state. Vigilantism was replaced by 'community policing', an institutionalised form of cooperation between the police and the population for achieving the legitimisation of the police.

Post-apartheid vigilante groups are often understood as the manifestation of a localised sovereignty that is active in the production of a moral community defined against certain groups (youth and foreigners, for example) that are perceived to represent a threat against this moral community (Buur 2006; Jensen 2005). I mentioned in Chapter 3 how the use of violence by vigilante groups, which was common against the youth of the township, was not only tolerated but was officially sanctioned by the state. Despite a massive transformation of the state's legal apparatus following democratisation, vigilantism appears as a historical tradition and a central feature of post-apartheid township life (Buur 2006; Fourchard 2011d; Kynoch 2005; Seekings and Lee 2002; Smith 2019; Super 2016, 2020). For Nicholas Rush Smith, the introduction of a strong procedural rights regime has enabling vigilantism rather than militating against it. Releasing without sanctioning a suspect known for his criminal activities in the township of KwaMashu (in Durban) often drive residents to take 'justice into their own hands' (Smith 2019). Residents in a context of high insecurity adopt their intervention in response to a perceived inefficiency and leniency of the criminal justice: banishment and forced exile of 'wrongdoers' from the township of Khayelitsha (in Cape Town) is an illicit practice which is locally perceived as an alternative to death (Super 2020). These examples show the importance of disaggregating and periodising vigilantism in each instance instead of conceptualising it as an unbroken continuation of the violence of the past (Buur and Jensen 2004, p. 143).

I argue after others that there is a partial continuity of former practices and major new ways of exercising local policing practices in the Cape Flats township between the late-apartheid and post-apartheid period. State community policing initiatives were established there to pacify vigilantism, to promote specific political party agendas, which in the process led to support financially a number of women in local security structures. In other words, vigilantism in the post-apartheid Cape Flats is shaped by three overlooked dimensions: limited pacification, uneven politicisation and feminisation of security largely due to its commodification.

Suggesting that vigilantism can be pacified by the police is clearly questionable, given that the South African police have engaged in widespread violence since the end of apartheid. South African journalists have reported the regular use of torture and a high number of extrajudicial executions at some of the country's police stations. Many of these killings took place in the criminal unit located in the Cato Manor neighbourhood of Durban and were covered up by police hierarchy.[25] When the police organise operations akin to raids against migrants and refugees, they are breaking the law; they destroy identity cards and engage in torture, intimidation and racketing; in so doing, they are helping to construct the image of the foreigner as a threat and entrenching xenophobic practices in the institution (Bruce 2007, p. 18; Demeestere 2016). Aggressive policing methods increased when Jacob Zuma came to power in 2009. An amendment to the criminal procedure authorising police officers to 'shoot to kill' (promulgated in 2012) strengthened the use of force by police officers and reinforced their relative impunity.[26] There has also been little pacification of the law enforcement techniques used against social or trade union movements: following the strike and subsequent massacre of 30 minors at the Marikana pithead in 2012, the commission of inquiry clearly stated that the strikers had no intention of attacking the police and that the deaths and injuries could have been avoided if they had adopted less brutal methods (Botiveau 2014). Eventually in dealing with vigilante groups themselves police agents help to generate violent activities through their discourse and practices or could be complicit with illicit practices or violence while extrajudicial killings by the police against black young men might also emulate similar practices by young townships dwellers (Owen and Cooper-Knock 2015; Smith 2019; Super 2020).

I argue however that stricter police supervision of the NW appears to be an integral part of the government's determination to ensure a police monopoly on violence, and that the police tolerate corporal punishments by vigilante groups only in some specific conditions.

The context of the post-apartheid Cape Flats well illustrate change and continuity in daily policing practices. As discussed in Chapter 3, this vast dormitory town populated by workers and employees of the Cape Town metropolis concentrates many of social problems inherited from apartheid (prevalent unemployment, drugs, gang fights). Criminality and violence are commonplace in everyday life in the Flats. An audit ordered in 2009 by the Department of Community Safety (DOCS) of Western Cape Province reveals that in the opinion of residents of Cape Flats, violent criminality and drug consumption had considerably increased since the end of apartheid.[27] In addition to the long-standing presence of gangs and the procession of score-settling incidents, the area has become a centre of widespread use of methaqualone drugs (known locally as *mandrax*), sold by prison and street gangs in the 1980s and 1990s, and amphetamine-based pills called *tik*, which have become widespread since 2000. According to the South African Medical Research Council, Cape Town had 200,000 users in 2008 and the municipality notes that the increase in crime and offences in the metropolis is largely linked to drug consumption.[28]. The consumption of *mandrax* and *tik* associated with Cape Town's coloured gang culture is mainly restricted to coloured populations where it affects all social milieus but more often young people, especially in the Cape Flats (Kapp 2008, pp. 193–194). In most neighbourhoods, it is considered the number one problem of local security groups.

To account for these dynamics, we have focused on two types of neighbourhoods with different profiles. Manenberg and Heideveld are archetypes of the former worker township dominated by several rival gangs. Today they are faced with a high unemployment rate (35%) and a high level of poverty (60% of households live on less than 160 euros per month, which is below the minimum subsistence level determined by Cape Town authorities in 2001). In contrast, the Mitchells Plain neighbourhoods (Rocklands, Westridge, Portland) has lower unemployment rates (below 15%) and a smaller percentage of poor households (less than 30%).[29] These neighbourhoods have a long tradition of vigilantism, which was perceptibly altered by a regime change in 1994. To observe the everyday changing practices I carried out an exhaustive examination of the local newspaper of Mitchells Plain, the *Plainsman*, for the period 1983–2010; I conducted numerous interviews with police officers in the field, Community Police Forum (CPF) presidents, leaders and members of a local NGO in Manenberg, and members of NW. I patrolled with seven different groups in 2009 and 2010 in Manenberg and Heideveld, and in Mitchells Plain (Rocklands, Eastridge, Westridge and the shopping centre) and I collected the life stories of women between 43 and 53 years of age involved in local security associations for more than 15 years.

Politicisation of Security Initiatives

The NWs were inspired by the experience of the Peacemakers (created in the mid-1970s and abolished in 1979), which had been based on the formation of volunteer patrols on a neighbourhood scale (Chapter 3). The first such groups were created in Mitchells Plain in the early 1980s and another a few years later in the townships of Manenberg, Heideveld and Surrey Estate (Jensen 2005, p. 229).[30] This movement had no connection with the UDF fight against apartheid (Seekings 2000). In fact, the relationship with the police during the 1980s was one of the main differences between the NWs and the organisations linked to anti-apartheid movements. Whereas the street and neighbourhood committees wanted to substitute a new political and social order for the old apartheid structures, NW groups were organised sometimes with and sometimes without the police, but seldom against them.

Thus in 1987, the police commissioner of Mitchells Plain vaunted 'the non-political, non-racial, and non-vigilante dimensions of the NW' and proposed to establish one in each neighbourhood.[31] By the end of the 1980s, the NW had become the dominant model in the area. An association called the Mitchells Plain Neighbourhood Watch Association (MPNWA) was set up to coordinate initiatives in the various neighbourhoods.[32] Its leaders and activists were mostly men, including Christian and Muslim leaders, members of ratepayers' associations, residents, clubs and NGOs, as well as social workers.

In a context marked by a very high degree of insecurity, guaranteeing protection became all the more crucial as the Cape Flats was at the centre of a recurring battle between the two main post-apartheid national political parties. The two sides were deeply divided, particularly in the Cape Province. Indeed, the province and the municipality are the only areas in the first two decades of post apartheid South Africa to have undergone several government changeovers between the main parties. At every election, the ANC competed in a close contest first with the National Party (1994–2000) and later the Democratic Alliance (DA) (since 2000) for the control of the main municipalities and the province.[33] As the DA rarely received votes in African neighbourhoods and the ANC had only a marginal presence in white municipalities, the two parties concentrated most of their efforts on the province's electoral majority, the coloureds especially in the Cape Flats.[34] The competition between the DA and the ANC significantly politicised the security initiatives following the end of apartheid, and was a feature of the fight against vigilantism, particularly the practices of the People against Gangsterism and Drugs (Pagad) during the second half of the 1990s.[35] Pagad's extremely violent operating methods – they claimed to have executed 30 drug dealers between 1998 and 1999 – (Kinnes 2000, p. 37), together with popular support for the organisation in those years, underscores the extent to which drug dealing had become a major concern for most residents. The support for vigilantism also shows that police efforts to combat drug traffic had been discredited. The Pagad movement was ultimately eliminated through police repression in 1999, but the problem did not go away, and security-oriented political activism continued to surface in the area.

Indeed, the initiatives of the police, the provincial government, and the municipality during this time and since reflected their fear of a new upsurge in vigilantism as well as recurrent conflict between the ANC and the DA. This was demonstrated by two major programmes. In 2001, the ANC provincial government launched the People-Orientated Problem Solving Policing Plan known as *Bambanani* ('stand shoulder to shoulder' in *Isixhosa*) in response to the popularity of the Pagad in the province. It mainly sought to train and pay a daily allowance of 50 South African Rands (5 euros) to hundreds of 'volunteers' to monitor sensitive locations such as trains, stations, schools, shopping areas and beaches.[36] The initiative was greeted with enthusiasm because it was the first time security volunteers were paid. The ANC clearly intended to popularise its actions in coloured neighbourhoods and to win the population over to the party through de facto competition with the long tradition of free NW service.

The revival of street committees is another example of the politicisation of community security initiatives. The groups, once symbols of the ANC anti-apartheid struggle in the 1980s, were reintroduced by Jeremy Vearey, the new director of the Mitchells Plain police force in 2007 and a charismatic ANC figure in the area. Unlike their original counterparts in the 1980s, the new committees were run by the police and aimed to mobilise the population against street-level drug traffic. They were set up to prevent the resurgence of violence involving Pagad-inspired methods that had taken place in certain neighbourhoods of Mitchells Plain in April 2007. Several drug dealers' houses had been burnt down by members of a local association called the Padlac (People Against Drugs, Liquor and Crime), supported by Helen Zille, mayor of Cape Town, president of the DA and head of the national opposition in South Africa. The street committees then became the focus of an ANC marketing operation that revived them on a countrywide scale to combat criminality, whereas the national opposition saw them primarily as an ANC effort to marginalise the NW.[37] It therefore seems obvious that the NW, the *Bambanani*, and the street committees were used by political parties to shore up or extend their local influence, as the head of community policing programmes for the Western Cape Province government, an active member of the ANC for 20 years, readily acknowledges.[38]

At the local level, the significance of the street committees and the NW was totally different. Even though the local leaders were members of the two parties, they deliberately depoliticised the initiatives ('the street committees are not an ANC programme', 'the NW do not belong to the DA') to legitimise operations that mobilised thousands of citizens every weekend to keep watch over the streets. For grassroots volunteers, opposing the NW to the street committees was pointless. Dozens of NW and street committee participants were not members of any political party and the reason for their commitment – repeated many times over during patrols – was that they had rid the neighbourhood of drug dealers. In many neighbourhoods, collaboration between the more longstanding, institutionalised NW and the more recent, less well equipped, and often spontaneous street committees was indeed the norm. Local security initiatives, which were part of national partisan struggles, were perceived by the base as dovetailing, overlapping or recycled schemes that were never really opposed to each other. Consequently, these initiatives cannot be seen as either simply community mobilisations or police initiatives or examples of political instrumentalisation: on the contrary, what characterised them all was the fact that they were co-produced by the police, the 'communities' and the political parties The process of bureaucratisation played a key role in their creation.

Limited Pacification and Bureaucratisation of Vigilantism

The literature has put considerable emphasis on the use of violence by vigilante groups in South Africa. In Cape Town, the NWs were seen as 'above the law', constructing their local power out of repeated violent acts, notably against young people (Jensen 2005, p. 237; Jensen 2008, pp. 184–192). My own experience of the patrols in Cape Flats partly attests to that violence, but the coercive aspect should not overshadow other dimensions that have seldom been discussed: the simultaneous appropriation of the new human rights rhetoric by vigilante members and the relative bureaucratisation of its organisations under the influence of partial state funding. This is in keeping with the wider bureaucratisation of society, which structures behaviours by norms, rules, procedures and formalities that is not only imposed from the top down, but which also requires the voluntary participation of individuals (Hibou 2013). In this process, individuals participating in vigilante groups conform to police procedures and requirements in terms of respecting human rights even if the police themselves could simultaneously abuse human rights.

The first point to be noted is that the use of violence has been gradually transformed within these organisations. The behaviour of the patrols in which I participated was rarely brutal or degrading, although there were exceptions: in one instance, some men caught with prostitutes at the beach were stripped naked in front of laughing vigilantes; in another, three youth trying to avoid a body search were slapped. During the seven nights I spent patrolling, there was only one noteworthy episode of violence: a youth accused by his cousin of attempted rape was seriously beaten up by a number of NW members with the help of the police. When the police arrived on the scene, they locked the culprit in their vehicle, as they discussed the case with the NW members and the victim. As she did not want to lay a charge against her cousin, they all concluded that the culprit won't be charged; they therefore authorised the various members of the watch to go inside and hit him as an instant form of justice, in that case to inflict a corporal punishment against the culprit for attempted rape. This practice enjoys the tacit approval of state representatives provided the corporal punishment remains invisible. A similar tendency has been observed in Port Elizabeth where corporal punishment was replaced by new legal norms, but nevertheless continued to take place in parallel, hidden locations (Buur 2005). Similarly, in situations where security has been delegated to street committees, the NW, or the courts in the African township of Kayelitsha in Cape Town, the police tolerate violence provided it stays at the local level; when the violence becomes too spectacular or known and involves a larger public, the state condemns it (Super 2016, pp. 476–477).

Since 1994, the dissemination of the human rights rhetoric has noticeably changed the discursive repertoire concerning the indispensable need for corporal punishment to educate youth, particularly in training workshops such as the West Coast Anti-Crime Forum created in 1996 from the merger of thirty community organisations in the province (Kinnes 2000). In the field, the use of violence against arrested individuals is impeded by a host of rules with which members willingly comply, including registering with an association authorised by the CPF, using walkie-talkies connected to police frequencies, and signing an attendance sheet at the police station before patrolling. In adjacent neighbourhoods, the intervention of street committees sometimes leads to corporal punishment, but also to banishing delinquents from the community and adopting the rhetoric of human and constitutional rights by the actors (Super 2016). A long description of my participatory observation in a Mitchells Plain patrol provides a clearer picture of the relationships between Watch members and residents. These relationships can by no means be reduced to their coercive dimension; compared with the daily practices of the apartheid period, they show breaks as well as continuities and engage many different forms of interaction between those patrolling and other residents in the course of a night patrol.

Friday 22 May 2009, 9 p.m. 75 people gather inside the walls of the Rocklands mosque, a well-off neighbourhood in Mitchells Plain made up of small, detached houses surrounded by little gardens comparable to the suburban communities of provincial cities in France. Half the group are men, half are women, between 40 and 70 years of age, most of them wage earners. Some are wealthy: two large four-wheel drive vehicles are provided free of charge for the patrol, others have torches, radios, bulletproof vests purchased at their own expense. Everyone is wearing a fluorescent yellow vest and a cap bearing the name of the association – the MPNWA – supplied by the Department of Community Safety (DOCS). They are all members of the association; three of them say they have been patrolling since the NW was set up in 1983, but four-fifths are newcomers who have been involved for just a few months. They work only on Friday and Saturday nights. Tonight, the 'chiefs' are present as well (perhaps because I asked to take part in the patrol): the police superintendent of the Rocklands police station, the president of the Mitchells Plain CPF, a local figure from the ANC, and a long-standing trade union leader. Two police officers and three reservists are there to lend a hand.

First, there are two prayers in Afrikaans, one said by an imam, the other by a priest, calling upon God to protect the patrol. A long-time member tells me this tradition dates back to the creation of the Watch in 1983: ecumenism vs. gangsterism. The patrol sets out immediately after the prayers. As the group moves through the streets, they receive encouragement and sometimes applause from residents: 'That's what we need in Rocklands, bravo, keep it up!' The young superintendent is delighted. He assumed his new duties only a fortnight ago: 'In the past, we could walk in groups of two or three police officers, but it wasn't enough to win over the community. Now we feel supported'. The president of the CPF adds: 'The police can count on one section of 300 Watch members in Rocklands'. Most of the members joined the association six months ago; they are fanatical about volunteering: 'You have to serve the community first and foremost'; 'You have to protect children and small children'; 'You have to help others'; 'We don't want to be paid in return'. They also reproduce the discourse of the police managers: 'We are the eyes and ears of the police'. Noting my surprise at the size of the group patrolling, Judy Williams, a member of the Watch for twelve years, qualifies the number: 'The NW has once again become very active due to the development of street committees that began two years ago in Mitchells Plain to try and eradicate drug traffic. The mobilisation around street committees revitalised the Rocklands NW, which had been dormant for several years'. She is wearing yellow *Bambanani* vest. She tells me: 'I used to be a member of the *Bambabani*: I patrolled the beach during the summer holidays for three years: I was paid by the DOCS'.

The first searches start with random encounters in the neighbourhood streets: they concern only young people, i.e. those who look young, those in groups or those that don't have children (fathers and mothers on their way home with their progeny were spared). Those who are stopped are familiar with the procedure and comply without complaint: they empty their pockets and put their cell phones, belongings, and caps on the pavement; the men search the boys, and the women search the girls, calmly. The Watch members are looking for two things: *tik* and screwdrivers, the most common, rudimentary weapon in the Flats and an indispensable tool for robbing cars. They actually find very few: 10 that particular night. Patrollers have no authority to interrogate people: only the police can do that, even though it is not an offence to carry a screwdriver. The object was confiscated, and the youths went on their way without being disturbed. Searches are accompanied by the same questions: 'Where are you going?', 'What are you doing here?', and the same advice: 'Go home', 'You have no business being outside at night', 'It's dangerous around here'. The young people acquiesce, move on, go home, or continue what they were doing, and we sometimes ran into them again later at night.

10 p.m.: all the Watch members gather in a large neighbourhood street in front of a petrol station. The group already seems smaller than before. The Public Relations Officer (PRO) of the neighbourhood police station has just arrived. He tells me that he is responsible for explaining the workings of the police and community initiatives to the population: his job consists mainly in writing articles for local

newspapers (the *Plainsman* and the *People's Post*) and Cape Town tabloids (*Daily Voice* and *The Sun*). His speech was well oiled: 'Street committees were used during the years of struggle and liberation, but people still remember them. They were revived two years ago in Mitchells Plains, and they were what mobilised the communities'. I take a photo of the police superintendent, and a photo of the Watch. 'Well, we're going to another neighbourhood, they need us elsewhere, too', the president of the CPF tells me as he leaves with the PRO, the superintendent, the two police officers and the three reservists. Once the 'chiefs' have left, a young volunteer grumbles: 'They act like they're doing us a favour by patrolling when they're being paid for it, whereas we are all volunteers'. Judy adds: 'That reservist there, I know him well: he used to be in the Watch, and now that he's a reservist; he's very happy with his uniform and his gun, but look at him, he obeys the police and no longer serves his community, that's the trouble with the reservists'. The volunteers start walking again; some of them go home. There are still at least a dozen of us in the group.

They decide to go and see three or four street committees in the neighbourhood. The residents of the street – men and women of all ages but mainly between 50 and 75 years of age – have lit a fire in front of the houses of drug dealers: they drink tea or coffee while monitoring the entrances and exits. I hear the same slogan: 'We are the eyes and ears of the police'. I express my surprise: 'But there is no one in these streets at this hour. There's nothing to watch'. 'That's precisely because we're here', I'm told. A Watch member adds: 'What is beautiful about our neighbourhood is that there is nobody loitering in the streets anymore'. We have a bite to eat. After a little while, a woman of about 50 tells me: 'I've lived in the neighbourhood for years, but I don't know my neighbours. These evenings and the weekend are a chance to meet people and have a nice time together. If a robber arrives, I go right home'. I can still hear the discourse of the PRO regarding the revived memory of street committees: 'So what is the relationship between these street committees and the ones in the 1980s', I ask. No one answers. The connection seems less obvious for the grassroots members of these committees. It is midnight. We move into relatively lively streets. Groups of young people stand around talking in the street. When we reach them, they all gather together on the steps to the garden of a house. They watch us go by with a smirk on their faces and only come back out into the street when the patrol is in the distance. Judy explains: 'We are not allowed to enter private property. They know that; it's a way for them to avoid being searched. Our powers as a Watch are limited'. A few streets further on, there is a big party going on in a garden, with loud music, dancing, laughter and cries of 'Bang *Bambanani*! Bang *Bambanani*!' as we go by. The Watch members are embarrassed and have a hard time telling me the young people are openly mocking them. The expression implies: 'The Bambanani are afraid of us young people', 'They are afraid to search us'.

This description shows continuity with the apartheid period – e.g. the curfew imposed on the young and the influence of adults within the organisations – but also significant breaks: violence is seldom used, the patrollers do not overstep what they are allowed to do, parents and grandparents lecture the young people but they do not want to be responsible for disciplining them. In all likelihood, the selective use of body searches on the young is a contemporary avatar of the excessive use of flogging under apartheid (Chapters 2 and 3). It should be noted that these young people are continually defying their parents' authority, either from the safe distance of a private space, which allows them to avoid the disciplinary space, or by pretending to go home. If the parents still think of the curfew as a method of social regulation, it has to be said that today's youth seldom observe it.

The transformation of relationships between generations through the intervention of legal limits on actions does not tell us everything about the mobilisation process, which has to be grasped starting with the initiative 'from below' all the way to its bureaucratic supervision. The residents who took part in the first violent actions against drug dealers in 2007 were transformed by the police into street committees in 2008, and then became the base for hiring new NW members, who in turn provided recruits for the South African reservists. Street mobilisation therefore does not take place today without bureaucratisation. The CPF managers and the police know that if they want to be able to count on the goodwill of unpaid volunteers, they have to maintain a high level of commitment beyond the initial mobilisation period. To do this, they rely on the culture of volunteering, which dates back to apartheid and has since been transformed into a commitment to a wide variety of causes.[39] In Mitchells Plain, the police organise monthly meetings during which quantitative objectives are assigned to each group of volunteers, usually for a short period of time. To encourage emulation and stimulate competition among the different neighbourhoods, every three months the president of the CPF and the managing director of the police ask neighbourhood representatives to announce how many new street committees and NW groups have been created, the number of new reserve forces recruited, and the type and number of offences and crimes recorded by each organisation.[40] 'The eyes and ears of the police' are also volunteer actors in the bureaucratisation of South African society in the townships.

Feminisation of Vigilantism

The motivation expressed by most members of the Mitchells Plain group (to combat delinquency and drug consumption) does not tell us who signs up for these organisations nor for how long or whether they are driven by their own local circumstances or the immediate environment. In reality, the degree of commitment differs from one neighbourhood to the next. It is also noteworthy that women are increasingly joining these traditionally male-dominated organisations. Many anthropologists and historians depict vigilante organisations as a minority – mainly of men – acting on behalf of the majority (Kirsch and Gratz 2010). Their action often comes in the wake of collective demonstrations, usually triggered by a murder, a robbery or a vendetta, as in the case of the street committees described above: once tighter control over the neighbourhood has brought about a return to 'normality', the groups grow weary and disband. Thus, the organisations often seem to be unified and homogeneous, with members striving to achieve the same objective, in this case to protect children from neighbourhood dangers.

There has been little careful research on the impacts of compensation or payment for joining an organisation or on the individual trajectories of those who join the movements, particularly those of women in these groups (see however Lar 2015; Nolte 2008; Revilla 2020, Smith 2019). Taking part in patrols in very poor neighbourhoods shows more clearly what type of reward the actors expect. An exploration of the biographies of vigilantes and the recent feminisation of the groups also reveals the crucial role played by funding grassroots initiatives in a context of widespread unemployment (Fourchard 2016). Thanks to the biographical approach, common in the sociology of activism, we can see that the organisational structure is not the only factor in determining the processes of mobilisation and demobilisation. The sequences of an individual's involvement in vigilante organisations may be consistent, as well as their continued commitment after periods of disaffiliation (Corrigall-Brown 2012).

The biographies of two women, Fazlin and Soraya, who joined the watch groups several years ago, attest to the participants' long-term personal commitment. They also suggest that instead of focusing on personal motives (which often results in overestimating people's initial motives for joining such groups), we can learn more from an overview of the successive moments of their involvement, especially about how the work changed its meaning for them and the interconnected factors that prompted their disengagement. From Fazlin's comments, it is easy to see the extent to which her security commitment has been closely connected to stages in her life and, in particular, to a family tragedy:

I was born in Athlone[41] on 12 July 1957 in a family of nine children. In 1974, the family moved to Manenberg. I went to primary and secondary school, and then left high school to work, first for my mother and later at a fish factory in the industrial zone of Epping. I had two sons and two daughters, and three of my children are still living. My second son was 18 when he died. It was in September 2001. After his death, I fell into a depression. Two friends came to see me and said: Fazlin, can we start an NW? I said yes. An NW had been operating in the Downs (the name of the neighbourhood) since 1993. At that time, I decided to join the NW to see what it was like. After my son's death, we set up a new NW in 2001.[42]

Soraya's account also sheds light on the important connections between social time and biographical time, and brings out the obvious – though rather ambiguous – influence of her family background:

I was born in 1967 in Deep Rivier [next to Retreat in Cape Town] in a family of four children. My father was an administrative secretary at the South African railway. The whole family moved to Manenberg in 1970. I went to primary school until grade 9. I began working in a factory when I was 18 in 1985. I worked there for sixteen years until 2000 when the factory closed. Then I became a cashier at a shop and, in 2006, a member of the NGO Proudly Manenberg. Now I am in charge of family conflict resolution in Manenberg for a research programme at the University of Western Cape. My father was part of the Peace-makers during the 1970s, but I made the decision on my own to start my NW in 1987. I was 20 years old. There was a lot of criminality in Manenberg, but I must admit that being a member of an NW was also a good way for a young girl my age to stay out at night [laughter]. I am a founding member of the NW. In 2001, my son was born and I decided to discontinue the NW. The NW ended because I was the main person in my neighbourhood encouraging others to patrol at night. In 2006, I became manager of a security area for Proudly Manenberg. I worked for the 'Safer Schools' project until 2009.[43]

These testimonials followed biographical timelines that are not necessarily consistent with the short timelines of the organisations and hence reveal continuities and the possibility of 'making a career' out of working in this field (Goffman 1968, p. 24). They are narratives of life experience, which describe something other than a need to protect young people. What they reveal is less a value system made up of desirable demands (Agrikoliansky 2001, p. 32) than an account of the contingent and highly personal process of individual commitment when faced with misfortune, the influence of neighbourhood encounters, and unspoken conjugal violence. Their commitment and disengagement are very closely linked to central episodes of family life (acquiring independence, having a child, losing a child) which are particularly decisive inasmuch as both Soraya and Fazlin are now single mothers, having rejected – or been abandoned by – a violent or absent husband.

In Cape Flats, state or community security organisations have provided an institutional framework that has structured the personal commitment of these women. In the mid-2000s, their initiatives tended to be determined by the policies of the provincial ANC government and its political ally, a local association named Proudly Manenberg. Those policies led to greater state control over the watch groups through partial funding. This is the context in which the transformation of the commitments of Fazlin and Soraya took place. Fazlin hints that those initiatives did not fundamentally affect her watch involvement, but that she supports them for the benefits she expects from them.

> When Ramatlakane, (the Minister of Community Safety in the Western Cape Province) arrived, he changed the NW into *Bambanani*. The *Bambanani* and the NW are the same thing. If you go to train for the *Bambanani*, you receive a tee-shirt, handcuffs, you get everything from the DOCS, the Department of Community Safety. They gave us bicycles. Ramatlakane compensated us for patrolling during the holiday period from 2005 to 2008. He also launched the 'Safer Schools' programme in 2005. We received an allowance of 50 Rands per volunteer and 60 for team leaders for each school.[44] We really loved Ramatlakane's initiative in the neighbourhood but the new DOCS had just stopped paying us compensation last year (2008). The patrolling of *shebeens*, parks, and all the public spaces during the holidays is now done by reservists.[45]

Taking part in a watch has become a way to gain access to meagre material resources or develop a network of relations. Depending on the governmental programme, NW members receive bicycles, jackets, tee shirts, radios and daily allowances for the sponsored programmes. Belonging to a local NGO therefore means being able to benefit from its leader's relationship with the ANC provincial government. It opens up the possibility of becoming a member of the 'Safer Schools' programme, which has been beneficial for all the schools in the neighbourhood.

The NGO security sector has 75 volunteers, including 70 women, mainly mothers between 30 and 60 years of age, most of them unemployed workers like Soraya or Fazlin. The commitment to the association is based on a moral contract between the volunteers and leaders: in exchange for recruitment by an NGO and an allowance paid for monitoring schools during the day, the members must carry out patrols on weekend nights.[46].

Soraya and Fazlin stand out from other female or male members by the length of their involvement in these watch organisations (20 years for Soraya, 15 for Fazlin), even though their participation has been fragmented during their lives. Their engagement differs from that of male wage earners, who participate in their NW only on Friday and Saturday evenings – and from the majority of enthusiastic volunteers whose commitment seldom lasts more than a few months. A small minority within the watch groups, they have made a career out of providing security on a neighbourhood scale, and as such have gradually become professionals. They have acquired expertise through their in-depth knowledge of the existing apparatus (of the police, neighbourhood organisations and political parties), the rights and duties of watch members, which actions are legal and which are illegal, and the opportunities available in their own neighbourhood or adjacent neighbourhoods. This long-term involvement has enabled both women to build networks of reciprocal local contacts that can be mobilised for more remunerative tasks: security at private evening receptions, remunerated staff at cultural events or at the World Cup in 2010.

Watch organisations have attracted more female members in recent years than law enforcement, notably the police.[47] My observations during patrols carried out in 2008 and 2009 show an increase in women members (between half and two-thirds were women), in all likelihood due to the partial funding method. Men avoid this type of poorly remunerated work and it carries little social status because it is not considered a 'real' profession, worthy of a man. For women, the daily allowance of 50 or 60 Rands is not enough to feed a family,[48] but it enables the majority of mothers to take care of their children after school, in a context in which women continue to bear the responsibility for their education and the burden of domestic tasks. The percentage of women NW members is higher in working class neighbourhoods heavily affected by unemployment.

The feminisation process has hardly affected NW practices, discourses or institutions. Fazlin, Soraya and the many women encountered during the patrols link their action to a conception of law enforcement rooted in the long history of the townships, which emphasises on giving priority to protecting children and youth from the dangers of the streets, conflicts between gangs, delinquency, and, in the past 20 years, drug consumption. Sheltering young people still implies a willingness to restrict their freedom. The participation of mothers in the patrols perpetuates a historic practice of closely associating delinquency with youth.

Conclusion

Once these groups of volunteers were replaced by security providers demanding payment for their services, local protection became a resource for a larger number of actors. This change, which took place in the poor areas of Ibadan and Lagos maybe around the 1980s, and more recently in the Cape Flats, clearly shows the unequal commodification of vigilantism.

In Ibadan, interviewees lamented the loss of a sense of belonging to a community with its rights, its duties and its forms of solidarity. The change is nevertheless associated with the loss of the charismatic power of the *bale*. The financial capability of households, although very low, has replaced elders' capability of supernatural powers. The individualisation and commodification of vigilantism has been substituted for the old system based on the responsibility of the neighbourhood chiefs to ensure everyone's protection. The commodification of security has thus opened up a 'market' in the poorest neighbourhoods abandoned by the industry of private security, which remains unaffordable for the vast majority of households in these neighbourhoods. The number of collective actors is still limited, however, and the residents are often forced to choose the least costly solution.[49] In contrast, the volunteer economy has in no way disappeared from Cape Flats, where an impressive number of parents police the neighbourhoods each week. Their volunteering stems from the associative tradition of the 1960s and 1970s and the political mobilisations of the 1980s, which since 1994 have turned into commitments to extremely varied causes including the enthusiasm for the NW or the street committees. Nevertheless, in pauperised working-class neighbourhoods, security now offers an opportunity for women to become professionals in the provision of paid services. In this instance, the commodification of vigilantism has been accompanied by feminisation of the sector in South Africa.

Commodification has not prevented greater regulation and bureaucratisation of protection in both countries. Those processes are obviously more developed in South Africa than in Nigeria. The short-term, teleological view of informalisation or privatisation of the state in Africa disregards the fact that the processes of institutionalisation and deinstitutionalisation are not linear. For decades, administrations, ministries and local police officers have looked favourably upon movements that provided assistance in areas where the police were absent and had long tolerated or even encouraged the use of violence. The banning of these organisations had very little to do with the fact that their violent practices challenged the sovereignty of the state. They have been suppressed, co-opted or institutionalised for political purposes, which governments deliberately confused with overriding public interests. The organisations were not banned or suppressed until they were identified as instruments of political enemies of the party in power (in 1952 the civic guards, in 1979 the Peacemakers, in 1999 the Pagad in South Africa, in 1966 the night guards and in 2002 the Bakassi Boys in Nigeria).

For most of the twentieth century, resorting to violence therefore became a way of regulating urban society in accordance with moral imperatives set by organisations that helped in large part to make violence an everyday occurrence. Today, those organisations are no longer authorised to dispense justice themselves, even if they continue to do so in one form or another. The Nigeria Police Force (NPF) monitors these areas remotely. The promotion of vigilantes has nevertheless introduced more state regulation than the literature on the new liberal urban order suggests: the police presence, if only episodic, is nevertheless real; vigilante organisations have more routine relations with the neighbourhood police station; they provide more detailed information on the offences that are committed, and when suspects are arrested, they are more regularly handed over to the NPF. All these elements testify to the emergence of an economy of vigilantism based on a technique of remote policing by the NPF combined with more individualised knowledge about the local population, an observation that needs to be confirmed by further empirical observations. In South Africa, this older political economy has been singularly strengthened since the end of apartheid by the bureaucratic participation of volunteers engaged in police work, supplying institutional police with a significant amount of information on crimes and offences committed in neighbourhoods. This economy is accompanied by the processes of co-opting and absorbing members of these organisations, but its subsequent trajectory has been very different in both countries: it has been based on the individual, voluntary commitment of South African reservists since the end of the 1960s. In Nigeria, the government has taken over collective organisations such as the NA Police between 1966 and 1970, vigilante state services in various Nigerian states between 1998 and 2002 and since 2007, the Nigeria Security and Civil Defence Corps.[50]

By identifying specific threats – youth and migrants in the Cape Flats, outsiders in the historical neighbourhood of Ibadan – these organisations have helped to build archetypal figures of exclusion that are not radically different from those produced by colonial engineering or apartheid. Along with young people, who have been discussed at length, temporary migrants are probably one of the other targets presented as a potentially threat to the cohesion of South African townships. During the political mobilisation of the 1980s and 1990s, this division, grounded in the institutionalisation of unequal access to the city and its advantages (work, housing, a decent life and leisure activities), led to extremely violent clashes between anti-apartheid militants from the townships, on the one hand, and migrants from hostels and protection gangs and rival parties, on the other (Sapire 1992; Sitas 1996). We can understand the xenophobic violence in South Africa during the 2000s by tracing the genealogy of the particular relationship that developed initially between urbans and temporary migrants, and later between urbans and foreign migrants. Similarly, the genealogy of the particular relationships between protection organisations that identified a threat with anyone from outside the neighbourhood as whatever came from outside will allow us to grasp the violence between native and non-native populations in Nigeria.

Notes

1. South Africa still has one of the highest murder rates in the world, even though it dropped from 66 per 100,000 inhabitants in 1994 (26,000 per year) to 35 per 100,000 inhabitants in 2020 (21,300 per year). Murder rates are often used as a basis for calculating international comparisons because they tend to be better reported to the police than other crimes and offences. Nevertheless, they are regularly a topic of controversy in South Africa: the figures vary by as much as 100% depending on the organisations, which leads to divergent interpretations by political parties, NGOs, journalists and security specialists.

2. Andrew Walker, 'Nigeria Police hold "Robber" Goat', BBC, 23 January 2009; interview with Andrew Walker, BBC correspondent in Abuja, Oxford, 20 May 2013, http://news.bbc.co.uk/2/hi/africa/7846822.stm

3. According to the figures provided by the Nigerian Federal Office of Statistics, the percentage of the country's poor population rose from 28% in 1980 to 65% in 1996, and the percentage of poor urbans rose from 3% in 1980 to 25 % in 1996. Federal Office of Statistics, *Poverty Profile for Nigeria: A Statistical Analysis of 1996–1997 National Consumer Survey (with Reference to 1980, 1985 and 1992 Surveys)*, Lagos, FOS, 1999, pp. 24–26.

4. 'Oyo Vigilantes launched', *Nigerian Tribune*, 7 April 1987.

5. Mobilisation Community Development Committee Edict, 1987, supplement of the *Oyo State of Nigeria Gazette*, 12 (14), 2 April 1987.

6. 'Oyo Vigilantes Launched', *Nigerian Tribune*, 7 April 1987.

7. Interview with Rasheed Aderinto, Ibadan, 22 January 2003.

8. Interview with Awe Latosa, Ibadan, 26 January 2003.

9. Interview with Oyeyemi Ajari, Alhadi G.O. Alatise, Ibadan, 23 January 2003.

10. Interview with Alahji Onitaru Alaodun, Ibadan, 29 January 2003.

11. Interview with Wahabi Alawepe Lawal, 14 January 2003, Osuelolale Oladineji, 7 February 2003, and Rasheed Aderinto, 22 January 2003, Ibadan.

12. 'We pay the police 100 *nairas* a month for the chairs we sit in when we have meetings with them and for repairs on their vehicles; the community even built a hall for them', interview with Wahabi Alawepe Lawal, Ibadan, 30 January 2003.

13. Interview with Wahabi Alawepe Lawal, Ibadan, 14 January 2003.

14. Interview with Lamidi Olawiwo, 21 January 2003 and Osuelolale Oladineji, 7 February 2003, Ibadan.

15. Interview with Ola Raufu, Ibadan, 12 February 2003.

16. The historic centre covers about one-third of the metropolitan area, which according to the most reliable estimates has three million inhabitants to date (Potts 2012, p. 1387).

17. Interview with Oyeyemi Ajari, Alhadi G.O. Alatise, Ibadan, 23 January 2003.

18. Interview with Amusa Adedapo, Ibadan, 29 January 2003.

19. Interviews with Oyeyemi Ajari, Alhadi G.O. Alatise, 23 January 2003 and Raliatu Adekanbi, Ibadan, 21 January 2003.

20. Interviews with Osuolala Oladineji, 7 January 2003 and with Alahji Onitaru Alaodun, 29 January 2003, Ibadan.

21. Interview with Raliatu Adekanbi, Ibadan, 21 January 2003.

22. Interview with Wahabi Alawepe Lawal, Ibadan, 30 January 2003.

23. Interview with Aladaji Alimi Buraimo, Ibadan, 28 January 2003.

24. See also the doctoral research under way of Lucie Revilla, *Pratiques de policing et acteurs locaux de l'ordre dans les quartiers populaires à Lagos et Khartoum*, Bordeaux, Sciences Po.

25. Journalists assert that 677 South Africans were killed by the police in 2014. Stephan Hofstatter and Mzilikazi wa Afrika, 'South Africa: Echoes of Apartheid, Investigating Allegations of Apartheid-Style Extrajudicial Killings in South Africa', *Al Jazeera*, 19 November 2015.

26. Pierre de Vos, 'Police Brutality Comes as a Surprise? Really?', *Daily Maverick*, 1 Marchs 2013.

27. Department of Community Safety, Western Cape Province, Safety Audits, KPMG, Le Cap, March 2009.

28. Janet Gie, *Crime in Cape Town, 2001–2008. A brief analysis of reported violent, property and drug related crime in Cape Town*, Cape Town, Strategic Development Information and GIS Department, January 2009.

29. Information and Knowledge Management Department (compiled by), *A population profile of Mitchells Plain. Socio-economic Information from the 2001 Census*, Cape Town, October 2005.

30. Interview with Ree Salomon and Malik Fajodien, founding members of the NW of Rockland, 22 May 2009; interview with Norman Jantjes, a social worker and NW promoter in Mitchells Plain in the early 1980s, 15 August 2009; interviews with Soraya Simpson and Abel Fazlin, founding members of the NW in Manenberg in 1987 and 1993, 20 January and 15 May 2009, Cape Town.

31. 'Crime Watch for "Plain", *Plainsman*, 15 July 1987; 'Crime Watches under spotlight', *Plainsman* 19 August 1987.

32. In 2009, the MPNWA claimed to have 2,000 paid-up members that patrolled their neighbourhoods on a voluntary basis. Interview with Michal Jacobs, president of the CPF of Mitchells Plain, Mitchells Plain police station, June 2009.

33. The ANC governed the Cape province between 1994 and 1998, then it was the National Party's turn from 1998 to 2001, before its control was given to a coalition of the New National Party and the ANC (2001–2004), then the ANC (2004–2009). During this decade, a new party, the Democratic Alliance (DA) resulting from the merger of the Democratic Party (DP) and the NNP in 2000, continued to increase its influence in the region. It was to govern the metropolis of Cape Town between 2000 and 2001, when the municipality came under the authority of an ANC–NP coalition, before once again returning to a coalition led by the DA in the local elections of 2006. The DA won the provincial elections in 2009, to date the only time the ANC has lost an election in South Africa.

34. According to the 2001 census, coloureds account for 46% of the total population in the metropolis of Cape Town, black Africans 31.2%; whites 21.1%; and Asian Indians 1.5%. (Dubresson and Jaglin 2009, p. 14).

35. See the conditions of this politicisation in Fourchard (2012).

36. *Bambanani Against Crime*, Le Cap, Western Cape Provincial Government, 2007.

37. 'Mbeki and Zuma Call for Street Committees', *Cape Argus*, 5 April 2008. Integrated Regional Information Networks (IRIN), *South Africa: Government Resurrecting Street Committees*, 24 September 2008; 'Enquiry to Probe Zille Allegations', *Plainsman*, 3 October 2007; 'I am Proud of where I come from. Director Vearey', *Plainsman*, 3 October 2007.

38. Interview with the director of the Community Police, Provincial Government, Cape Town, August 2009.

39. These causes include the fight against violence and drugs, releasing gangsters on bail, domestic violence and child abuse, and AIDS and in favour of helping the elderly and orphans and promoting world sports events like the World Football Cup in 2010.

40. Meeting of the Mitchells Plain CPF, June 2009. Central police station in Mitchells Plain.

41. A coloured, predominantly middle-class neighbourhood in Cape Town.

42. Interview with Fazlin, Manenberg, Cape Town, 15 January 2009.

43. Interview with Soraya in February May and June 2009 and September 2010.

44. 5 or 6 Euros, respectively.

45. Interview with Fazlin, Manenberg, Cape Town, 15 January 2009.

46. Interview with Mario Wanza, leader of Proudly, Manenberg, March 2009.

47. The police began recruiting women late – women were not offered posts until 1972 – and the process of feminising the police ranks has been slow: women accounted for 20% of South African police officers in 1996, 34% in 2014. There has been significant progress; however, especially among reservists, even though the statistics are incomplete (45% in the province of Gauteng in 2013) (Hendricks and Musavengana 2010, p. 131; (Forster-Towne, 2013, p. 29). South African Police Service, *Annual Report, 2013–2014*, Pretoria, 2014, p. 1.

48. Equivalent to about 100 euros per month.

49. Contrary to the argument put forward by Bruce Baker that in most African countries people can choose from among a number of protection systems (Baker 2008).

50. The surveillance organisations that monitored the thicket of pipelines against threats of piracy were absorbed by the federal state in 2007, becoming the Nigeria Security and Civil Defence Corps in 2007, one of the federal government's largest paramilitary corps. This 40,000-member corps ensure the security of the country's principal infrastructures, including petrol stations during a period of scarcity and secondary schools in areas menaced by Boko Haram in the Northeast of the country.

PART III

Politics of the Street, Politics in the Office

Part Three analyses power relationships at the micro level of streets, bus terminals referred to as 'motor parks' and local government offices between individuals in positions of authority (political leaders, civil servants, trade union members) and a host of subordinate actors (bus drivers, tax collectors, unemployed workers, ordinary citizen seeking for a document) in three main metropolises of Nigeria (Lagos, Ibadan and Jos)[1]. Chapter 5 explores how the political leaders of Lagos and Ibadan have used their networks to provide infrastructures and services to the population. It looks more specifically at one of the most powerful unions in Nigeria, the National Union of Road Transport Workers (NURTW), which governs motor parks and collects 'taxes' or 'fees' over millions of drivers with the support of the political class in power. These revenues are of the most important illegal and unofficial taxation organised on a daily routine in Nigeria: illegal as levying taxes in parks is according to the constitution the sole responsibility of local governments and unofficial as the sums collected are largely non-receipted. Chapter 6 interrogates interactions between civil servants and citizens seeking a certificate of *indigene* in local government front offices in the city of Ibadan, Oyo state capital, 100 km north of Lagos and of Jos, Plateau state capital in the central part of the country. A quota system has been put in place in all federal administrations starting in the 1980s, and guarantees today the representation of citizens from the 36 states and the 774 local governments of the country. Under this principle, all people claiming to be *indigenes* of a state are guaranteed certain rights (access to civil service positions and political mandates, reduced fees in universities), whereas *non-indigenes* are *de facto* excluded from those same rights. Access to those resources and services is thus partly conditional upon an individual's *indigene* origin, which must be certified by the local government. Certificates of *indigene* embody official and illegal discrimination in Nigeria: it discriminates citizens based on their origin, which is illegal according to the constitution, but it is official, in the original meaning of the word, as certificates are issued in the *office* of local governments all over the country. The level of violence linked to the control of parks or to indigeneity, a common term employed by actors and researchers to describe *indigene/non indigene* divide in Nigeria – is astonishingly dramatic. For instance, in September 2001, the contested nomination of a civil servant of *non-indigene* origin in one local government position in Jos triggered a week of mass violence (death of more than 1000 people, destruction of hundreds of houses, schools, shops, mosques and churches). The following month, a major fight between factions of the NURTW in Ibadan reported probably 300 deaths in two days. The scale of massacres appears to be unique in the history of the two cities (Albert 2007, p. 142; Higazi 2007; Madueke and Vermeulen 2018, p. 39). They were however followed by a number of clashes with often similar levels of violence in the following years. In sum, motor parks and local governments are privilege places of observation of both ordinary social practices and mass violence, of exercise of power through taxation or discrimination. This last part wishes to explore the ways in which these places become politicised or not, might generate violence and exclude as much as include citizens.

It is first far from obvious that neither NURTW members nor civil servants in local governments constituted what is referred to by Pierre Bourdieu as a 'political field', or 'a microcosm relatively autonomous within the larger social word' (Bourdieu 2000, p. 52). Local transport union members working in the street and bureaucrats working in front office are part of a larger web of relationships dominated by more influential civil servants, union leaders and politicians from whom they cannot easily become autonomous, as these two chapters show, even if their role also reshapes the boundaries of the local political landscape. Riots, violent or more mundane demonstrations in the motor parks or against exclusion of *non-indigene* are not part of a social movement denouncing the injustice of the system or contesting state policies and even less a form of political insurgency producing emancipatory sequence or new forms of democratisation (Dikec and Swyngedouw 2017). The various episodes of mass violence tend instead to reinforce the position of dominant actors: certain top union leaders in motor parks and *pro indigene* politicians in local government offices. It is also questionable whether these contestations are a demand of equality or what is referred to as *le politique* by Jacques Rancière (Rancière 1998, pp. 113–115). There is no mobilisation at local or national level of transport users or drivers to protest against NURTW illegal taxation in motor parks despite the fact that a majority of drivers see this as extortion (Agbiboa 2018, 2020). There is no contestation by citizens or activists of the principle of indigeneity perceived instead as stabilising since political and bureaucratic positions are distributed all over Nigeria according to quotas (Fourchard 2015). Thus, conflicts and protestations are not against the very foundation of exclusion. Actually, a national system that excludes some citizens of their fundamental rights because of their origin does not explain why in some places it is highly inflammable while in others it produces an apparent form of consensus. This last part looks at politics beyond state policies and its contestation (central in classical urban politics and for a number of Marxist scholars) and beyond a 'post political' analysis in which politics are reduced to 'good governance', 'new public management' and 'best practices' – an approach antagonized to a prophetic vision of politics expressed only in its revolutionary moment (Le Galès 2020).

In other words, what is happening in these places does not easily fit what a number of scholars would qualify as 'political' in its emancipatory dimensions. Politics is instead understood as participating in power. With his brief formula, Weber insists that striving for power is a necessary condition for acting politically (Weber 1919, p. 36). Power, in Weber's nominalistic view of politics, consists only of the shares and their distribution (Palonen 2003, p. 173). In the exercise of power, the term *eigensinn*, developed by Alf Lüdtke (2000, 2015a) accounts for the ambivalence of actors by revealing the appropriation and mundane uses of structures of domination. *Eigensinn* refers to the multiple ways individuals relate to power, from resistance to hyper-obedience, from the determination to perform better than what it is asked for to the desire to participate in power. The notion of *eigensinn* shows how individual appropriation of the structures of domination helps to legitimise those structures. For citizens in situations involving the practical exercise of power, this translates into the desire to have a normal life, receive social or economic benefits, be assured of a minimum standard of living and have a mastery or at least the feeling of mastering a system (Hibou 2011, pp. 26–33). Participating in power does not necessarily mean politicisation. In Weber's conceptual horizon, while policy refers to the regulating aspect of politics, politicisation marks an opening of something as political, as 'playable': 'politicization either introduces new items, which alters the relationships between the existing ones, or dismisses existing items' (Palonen 2003, p. 184). In practical terms, politics of the street looks at what is happening in motor parks and bus terminals that are physically located in the street because of the activities of transporting people and collecting money. Politics of the street designates by extension the transformation of public space (including the residence of influential political leaders) into places of protestations, violence and mobilisations, a process qualified as the politicisation of urban spaces (chap 5). Politics in the office explores the everyday working of issuing certificates of indigenes in local government offices as the manifestation of a new urban politics of exclusion. For Weber, the combination of written documents and a continuous operation by officials constitutes the 'office' (*Bureau*) (Weber 1978, p. 219). Documents are not however the passive instruments of bureaucratic organisations formed though norms and rules as described by

Weber but rather are constitutive of bureaucratic activities and the social relations formed through them (Hull 2012b, pp. 18–19). Exploring patronage networks in motor parks and other public places (Chapter 5) and bureaucrat-users' interfaces in local governments (Chapter 6) are aimed to understand the boundaries between those to exclude and those to include, the very concrete experiences of violence and uneven forms of politicisation triggered by such relationships. Patronage, clientelism and bureaucracy are here fundamental notions through which these experiences are analysed.

Patronage relationships have been at the centre of reflections on the nature, functions, scope and formation of the contemporary state in Africa in general and Nigeria in particular (Bayart 1989; Joseph 1987; Médard 1991). Patronage is a type of clientelism that consists – for the patron – in using state resources to provide work and services for political clients (van de Walle 2007, p. 51). The patrons are often politicians who distribute state funds to local leaders, well-known personalities in business, trade unions and associations. For the client, putting oneself at the service of a patron means negotiating the terms of one's inclusion in a network based on a set of mutual social obligations linking the parties to each other (Pitcher, Moran and Johnson 2009). According to Nicolas van de Walle, clientelism has historically taken the form of prebendal politics in Africa, also called *prebendalism* (van de Walle 2014, p. 237), a notion that has become widespread in Nigeria since it was used by Richard Joseph (1987) to describe the democratic experience of the Second Republic (1979–1983). Political clientelism is more restrictive as it is held to be the selective distribution of goods and services by politicians to benefit voters in exchange for their political loyalty (van de Walle 2014, p. 232) while patronage includes a more diverse range of actors: patrons are also 'big men' at the head of social networks that 'connect' their clients with bureaucrats and political leaders (Bayart 1989, p. 217). Patronage and clientelism are not specific forms of deviance characteristic of underdeveloped authoritarian regimes; they are not anachronistic relics that disappeared with the advent of urbanisation, increased education and the expansion of the middle class (Mattina 2016) but are rather a particular way of representing interests (Briquet and Sawicki 1998) distinct from corruption.

I will use the term patronage according to three general acceptations widespread in Nigeria. Patrons are first professional politicians working in office elected or nominated for one or several mandates. Nigeria is constituted of 36 states and 774 local governments and governors are considered to be patrons of their state like chairmen patrons of their local government areas. Some of them have a determinant influence even after leaving office: no governor in Lagos could be elected without the support of Bola Tinubu (1999–2007), the first Lagos state governor elected since the return of a civilian regime in 1999. Patrons are secondly politicians of the street: they are heads of social networks who govern from their private residence or from the public space (streets, motor parks, markets) without being elected and without an official mandate. Those referred to as godfathers (because they finance gubernatorial election campaigns) play a fundamental role in providing daily services and work to tens of thousands of dependants, triggering violence during elections, or paying electricity bills in neighbourhoods like Lamidi Adedibu, the most influential politician in Ibadan between 1999 and his death in 2008 (Chapters 5 and 6). Patrons are not limited to the 'big men' in politics. The third and last acceptation refers to persons as varied as civil servants, property owners, union leaders or office workers, in other words, individuals able to maintain a more or less sizeable network of clients, friends and relatives and invested of a portion of authority (Barber 1991; Barnes 1986). Transport union local leaders and civil servants working in local governments obey a superior authority but their power to collect 'fees' and grant services or authorisation in the street or in office give them an important level of discretion in implementing rules or playing politics.

The body of work on contemporary Nigerian politics remains, with a few exceptions, a view from the top of the government or from the political elite. There is a need to explore more mundane forms of politics (in office and in the street) and to understand the process of uneven politicisation in looking at the interplay between politicians, bureaucrats and their networks in the city. This is inspired by a set of urban sociological works. Auyero explores everyday dealings of political brokers, the practices and perspectives of 'clients' and the problem-solving network that links 'clients,' brokers and political 'patrons' (Auyero 2000, 58) that help to rethink how subaltern actors are politicised through a variety of means. Similarly, poor residents at the outskirts of Cairo are politicised by 'intermediaries' (gang leaders, prominent citizens, Islamists) who attempt to integrate the party structure and distribute a portion of state funds to the people living in their immediate vicinity (Haenni 2005). Poor urban groups in India occupy land, pressure municipal and state administrations to transform them into multiple *de facto* tenures, a process referred to as 'occupancy urbanism' (Solomon 2018). And at the heart of the daily negotiations and conflicts between poor citizens and local authorities in Middle East cities, one finds what Assef Bayat calls the 'quiet encroachment of the ordinary', namely, fragmented mobilisations extended by the periodic efforts made by families or individuals to access services or collective resources (land, housing, utilities) and urban spaces (streets, marketplaces, parking)

(Bayat 1997, 2000). Following such an approach allows us to reassess the ambivalence of vertical ties between urban residents, street-level bureaucrats, intermediaries and politicians that can bring opportunities for political inclusion, foster public action, help to extend service deliveries and lead to violent mobilisations.

Note

1. I had to abandon a comparative analysis with South African metropolises for want of first-hand sources.

CHAPTER 5

Patronage, Taxation and the Politicisation of Urban Space

In the early 2000s, Lagos was composed of a tiny elite concentrating on enormous wealth and millions of residents working in the so-called 'unregulated informal economy': any urban space available (streets, roads, highways, parks, roundabouts, underbridge, motor parks) were turned into an open air market that absorbs a continual influx of newcomers (Packer 2006). For Rem Koolhaas and his Harvard colleagues, Lagos was then 'the paradigm and the extreme, pathological form of the West African city' (Koolhaas et al. 2000, p. 652; Koolhaas 2002, p. 183), an 'icon of West African urbanity [that] inverts every essential characteristic of the so-called modern city' (quoted in Enwezor 2002, p. 113). Despite the lack of all the basic amenities and public services, Lagos continues to function as a city because it is conceived as a series of self-regulatory systems that has freed itself from the constraint of planning and public intervention. Lagos illustrates the large-scale efficacy of systems and agents considered marginal, informal or illegal. Scholars across the world were enthusiastic about this new celebration of informality and Koolhaas's idea that Lagos was the future of the urban world. Others contested an essentialist vision of the 'African city' that put an exaggerated emphasis on its exceptional character (Gandy 2006, p. 390) while ignoring the influence of bureaucratic and political relationships that weigh on the everyday life of 'informal' actors (Fourchard 2011b).

In 2020, the largest city of the continent is considered one of its best managed metropolises. Bureaucratic reforms have increased the Lagos budget by 20 times in only 15 years (Cheeseman and de Gramont 2017; de Gramont 2015) making Lagos home to the single-most impressive 'tax turnaround' in recent African history (Moore, Prochard and Fjeldstard 2008). Lagos state governors have implemented consistent public policies in terms of transport, waste management, road construction, recovering and greening

Classify, Exclude, Police: Urban Lives in South Africa and Nigeria, English Language First Edition. Laurent Fourchard.
© 2021 John Wiley & Sons Ltd. Published 2021 by John Wiley & Sons Ltd.

of public space, revitalisation of central neighbourhoods and building tax-free zones to attract foreign investments (de Gramont 2015). Providing infrastructures and services have helped create the impression that the government is delivering on its promises, thereby facilitating compliance to pay taxes (Bodea and LeBas 2016; Owen and Goodfellow 2018). The success of Lagos state was so impressive that the Governors' Forum of Nigeria considered it as a 'governance model' for other Nigerian states (Roelofs 2016, p. 117).

Replicating this new 'model' is far from easy, however. At only 100 kilometres north of Lagos, Ibadan, the capital of Oyo state, the second largest city in Southern Nigeria with 3–5 million inhabitants, has actually witnessed a radically different post military trajectory. Despite sharing a number of similar conditions with Lagos state (an educated elite, a strong business community, state governors belonging to the same political family for a number of years, an existing fiscal basis), Oyo state government was rated in 2005 as one of the lowest-performing of the 36 states in the country (Roelofs 2016, p. 107). This has hardly changed since then. Compared with the citizens of ten other Nigerian cities, Ibadan residents have the lowest compliance to pay taxes and engage instead in 'self-help' provision of goods without the assistance of the state (Bodea and LeBas 2016). To a large extent, Ibadan looks like the rest of Nigeria: since the restoration of democracy in 1999, any breaks from past political practices are thought to be either modest or merely to have eliminated previous patronage logic (Adebanwi and Obadare 2013; Obadare 2007) despite an attempt to replicate the 'Lagos model' in Ibadan by the Oyo state governor between 2011 and 2019 (Roelofs 2019). Moreover, despite its recent successes, Lagos has not erased his past: patronage and clientelism coexist with bureaucratic efficiency (Joseph 2013; Nolte and Hoffmann 2013) while tax compliance does not mean that 'informal taxes' levied by non-state actors have disappeared (Agbiboa 2020; de Gramont 2015).

The relations between economic actors and state agents commonly labelled the 'informal economy' did not help much to explore patronage and urban studies in Africa. This notion has several connotations and it has acquired radically different political meanings since it was invented in the 1970s but when authors use the expression, it usually refers to self-employment, i.e. either non-wage work or activities that are not subject to taxation or state regulation. This is not the place to re-examine the criticisms concerning the vacuity of the dualism between the formal and the informal, and how they are interwoven in the production of the social world. Informality has probably become a mystical notion conveying too multiple readings (Haid and Hildbrandt 2019). We would simply note that many studies on urban informality have too often emphasised the vitality of horizontal ties and tended to overlook vertical relations. Less has been said on the merging of political party and local institutions in providing services or urban infrastructures (Bénit-Gbaffou et al. 2013; Ginisty and Vivet 2012; Simone 2004), on hierarchical social relations working within networks of informal actors (Lindell and Utas 2012), on gatekeepers controlling access to urban resources or mediating relations between residents and politicians (Salvaire 2019; Utas 2012).

Interestingly patronage politics, the transport sectors and elections have been an emerging academic focus in Nigeria (Agbiboa 2020; Albert 2007; Fourchard 2011b; Obadare 2007; Omobowale and Fayiga 2017), a connection also seen in other contexts (Cissokho 2016; Titeca 2014). Empirical research on the everyday modes of operating of transport unions and associations in Nigeria and other African countries is instead still limited. There is much to understand on the porous border between legal and illegal and state and non-state actors. In Cape Town, transport members act as a quasi-mafia and violate the law to produce forms of governance in conditions of extreme precariousness (Bähre 2014). Minibus transports in Nairobi perceived as thugs since the 1990s (Mutongi 2006) might work 'outside the law' in association with militia groups providing security in exchange of 'taxes' (Rasmussen 2012). In Nigeria, despite, the pioneering work of Albert (2007), Agbiboa (2018) rightly notes that 'we still know surprisingly little about the intermediary role and micro-level dynamics of semiformal transport unionists collecting taxes in motor-parks and bus stops in virtually every African city'.

This chapter is based on a 20-year observation of Lagos and Ibadan politics, tens of interviews with politicians and bureaucrats especially in local governments in Lagos (Bariga and Surelere, Yaba, Lagos Island) and Ibadan (Ibadan North, Akinyele). It is also based on 40 interviews of local union leaders and members in three motor parks in Ibadan (Akinyele, Ojoo, Sango) and seven in Lagos (Bariga, Makoko, Obalande, Oshodi, Oyimgbo, Pako Aguda, Yaba) in January 2018 and January 2020. The chapter first looks at the role of the two first governors in Lagos and of the most influential godfather in the capital of Oyo state, Ibadan, since the return of democracy in 1999. The implementation of a metropolitan project in Lagos indicates the ability of state governors to control federal allocation revenues and the urban revenues through a very broad network of patrons, clients and an expanding bureaucracy, while in Ibadan, ongoing competition amongst political patrons has exhausted state resources without leading to any urban or planning development. Beyond these narratives of successes and failures the second part looks how bus terminals are both central places of illegal tax collection and major places of politicisation and conflicts. Uncertain and unwritten arrangements between union members, political patrons, government officials and police officers are key to understanding the government with regard to transport and episodes of violence, the extortion of money from drivers and the inclusion within the union of tens of thousands of marginalised and poor men. Focusing on in-between positions – i.e. locally based politicians, local union leaders who act as quasi-street-level bureaucrats – helps to understand their bargaining or discretionary power, to investigate how they ignore, navigate or make use of established political networks and the organised collection of 'fees' and the sharing of this revenue among a set of institutional actors.

Patronage and Urban Projects

Matthew Gandy presented what has been called the 'infrastructure crisis' of the Nigerian capital as the product of authoritarian colonial and military regimes, the corruption of political elites and of the structural adjustment plan and the deindustrialisation of the 1980s and 1990s (Gandy 2006). All these points have to be taken into account, but the abandonment of basic services and the absence of large-scale projects are also due to opposition between the federal government and the political leaders of Lagos, which became the seat of two rival powers in the second half of the twentieth century. On the one hand, the capital has housed first the colonial and then the federal government (1914–1991), dominated by a coalition of political parties of the North and the East during civilian rule (1954–1960, 1960–1983) and by a clique of Northern generals during the military regimes (1966–1979, 1983–1999). On the other hand, the Action Group (AG), the dominant party in the Western Region in Lagos and Ibadan, and its leader, Obafemi Awolowo (1909–1987), have always been in the opposition to the federal government and have never been able to win a national election. Awolowo is an iconic figure in the South-West of Nigeria. He took over the idea of the nation developed by local intellectuals and built the Action Group starting in the 1950s, relying on networks rooted in local life, celebrating Yoruba cultural nationalism and insisting on education for all, social solidarity, and basic infrastructure as his political programme (Nolte 2009). Popular support for these projects regularly ensured the victory of the AG (and its successors) in local elections (from 1955 to 1966), elections of the governor of Lagos state and Oyo state during the Second Republic (1979–1983) and since civilian government was restored in 1999. The long-standing antagonism between the federal government and the Awoist tradition has been marked by recurring conflicts over the distribution of federal revenue, which is briefly summarised here.

During the Biafran War, the government launched a process of fiscal centralisation: taxes on imports and exports were no longer given back to the states, but instead were transferred to the federal budget. Oil-producing states that were still receiving 50% of oil revenues in 1969 saw their percentage drop to 1% under Sani Abacha (1993–1998). Fiscal centralisation bolstered the weight of federal patronage and the custodian state at the expense of the federated states. Henceforth, the generals in power (1983–1999) resorted to highly selective revenue distribution in the context of economic recession (1983–1999), the Structural Adjustment Programme (from 1986 onwards), and the reduction of the federal state budget. When Lagos, the former federal capital, and Ibadan, a former regional capital, became the capitals of Lagos state (1967) and Oyo state (1976), they had no investment plans and their annual budgets depended mainly on federal allocations managed by military governors who were not accountable to the population. The subsequent 'infrastructure crisis' in both metropolises was first and foremost the result of the federal state's fiscal centralism and the prebendalism of the governors and presidents of local governments. It has to be understood in light of the federal government's decision to invest massively – during a period of budget restriction in the 1980s and 1990s – in the construction of the ultramodern capital of Abuja.

In other words, during thirty years of military regimes, the centralisation of revenues gave the federal government the ability to subordinate numerous political actors. The fiscal, demographic and economic influence of the initial three large regions gave way to 36 states, all of which were very dependent on the central government. This hyper-centralisation of the tax system has not changed since the introduction of a civilian government in 1999.[1] The distribution of federal allocations has remained the same: the federal government reserves more than half of the revenue for itself; the 36 states and 776 local governments share the other half.[2] Three changes have however shaped differently the two major cities of the South since the return to democracy in 1999.

First, the regular payment of federal allocations to the states enabled the creation of patronage networks outside the People's Democratic Party (PDP), the party in power at the federal level between 1999 and 2015. This is the case of the states in the south-west, which still appear to be bastions of the opposition, heir to the Awoist tradition. Lagos state has been the leader of this opposition since 1999 under various labels (Alliance for Democracy (AD) in 1999; Action Congress (AC) in 2002; Action Congress of Nigeria (ACN) in 2008; All Progressive Congress (APC) since 2013). Lagos is governed by this same political family since 1999. The main patron is the first governor of Lagos state, Bola Tinubu (1999–2007), in the sense that no governor could be elected without his support. This political stability has helped to implement a reformist agenda on the long term (Cheeseman and de Gramont

2017). Ibadan by contrast has witnessed important and regular change in power from AD (1999–2003) to PDP (2003–2011) to ACN (2011–2019) and back to PDP since 2019. In the Nigeria context, this has brought political instability, lack of continuity in public policies and the absence a long-term project for the city.

Second, Lagos state governments were able to increase taxes and turned them into service delivery, which was not the case in Ibadan. The revenues of the states have considerably increased since 2000, in large part due to the rise in the price of crude oil: the budget of Lagos state was multiplied by 15 between 2000 and 2016, and that of Oyo state by 4 between 2005 and 2016.[3] More than 80% of the budget for Oyo state depends today on federal allocations, as it did before 1999. The percentage earmarked for Lagos has dropped significantly however: in 2015, it accounted for only 25% of the total state budget, with 75% coming from internal generated revenue, mainly through taxes on the personal income taxes on wage workers and generated by the metropolitan economy. Revenue once federal is now urban, and it is a major instrument of the metropolitan project.

Finally, with the return of electoral politics, governors and godfathers in Lagos state and Oyo state have integrated, won over and increased a number of influential actors in their networks. In Lagos, the two first elected governors succeeded in transforming the state departments, the different agencies and the old and new local governments into a larger bureaucratic and political network working towards the metropolitan project. In Ibadan instead different antagonist governors have favoured peculiar networks and heightened partisan competition in the city, which has led to numerous clashes and used up state resources.

The Amala Politics in Ibadan

The evolution of Oyo state has been similar to that of many other states in Nigeria since 1999. Patronage and corruption predominate in politics, modelled on the system that prevailed in the preceding decades: despite a fourfold budget increase in 10 years, infrastructures (roads, sanitation and educational facilities) and basic services (access to potable water and electricity) have remained practically unchanged. The political situation is particularly volatile and the governors in place have constantly had to deal with the state's most prominent political godfathers, notably Alhaji Lamidi Adedibu (1924–2008), a businessman who established himself as one of the chief political actors since the return to democracy. He played a key role in the violence linked to motor parks and during electoral campaigns, but he enjoyed popularity and prestige from the forms of patronage he had succeeded in developing until he died.

Born into an Ibadan family, Adedibu went into politics during the 1950s, becoming an active AG militant until the coup d'état in 1966. In 1979, he switched to the National Party of Nigeria (NPN), where he benefited from federal government support under three presidents: Shehu Shagari (1979–1983), Ibrahim Babangida (1985–1993) and Sani Abacha (1993–1998) (Agbaje 2003, p. 10). During the years of structural adjustment, marked by a sharp decline in wage labour and cacao cultivation as well as widespread impoverishment, it became imperative to belong to an extended family or develop other personal networks (Agbaje 2003; Cooper 2002, p. 125). Adedibu widened increased his social base by distributing meals and money, and paying for the services, school fees and Dawa activities[4] of many unemployed persons, orphans, vagrants, hauliers, Koranic schoolteachers and imams. These individuals gathered in Adedibu's huge compound in the Molete neighbourhood in the centre of Ibadan, where people trying to survive or attending to their business converged with candidates for electoral positions who wanted to capture Ibadan's political market (Agbaje 2003, p. 17; Obadare 2007, pp. 117–118). The compound became a space for the production of power. One of the key moments was at mealtime, when those present shared a meal of *amala* (a popular yam-based dish in the south-west) after Friday night prayers at the mosque adjacent to the compound. Loyalty, allegiance and inclusion in political networks were built around this ritual of table-companionship and politics of the belly (Bayart 1989). Known as 'amala politics', the politics of the belly is part of the long history of redistribution by political leaders to their clients or relatives in Ibadan.[5]

In 2000, President Olusegun Obasanjo decided to support Adedibu to win the state over to the PDP, which immediately mobilised its networks and managed to have two PDP governors elected in succession: Rasheed Ladoja in 2003 and Christopher Adebayo Alao-Akala in 2007. He acquired his status as kingmaker or godfather at that point, taking advantage of the redistribution of federal revenue on a local scale, his own influence at the head of multiple networks, his direct link to Obasanjo and his acquaintances in the police and state administration. Clearly, buying votes directly is not the most efficient method, and the few studies that exist, including those on Nigeria, suggest that it is not given priority by parties or patrons (Banégas 1998, pp. 78–79; van de Walle 2007, p. 64). Violence and cheating during the elections and the inclusion of numerous subaltern actors in political networks are certainly more decisive mechanisms. Adedibu is known to have recruited thugs before and during elections, who specialised in violence, intimidation and fraud, to eliminate opponents, cause mayhem at their meetings, ransack hostile radio stations and 'monitor' and/or stuff ballot boxes. In some instances, the thugs were groups of youth already engaging in a particular activity. In 2003, for example, he hired a group involved in selling cannabis in the Mapo Hall neighbourhood, which consequently received his protection against police arrest. He offered another group of young boxers from the Sabo-Mokola neighbourhood 5,000 nairas per week and a daily meal at his residence. In both cases, their politicisation did not entail joining the party (they did not have PDP cards and did not attend their meetings) but pertained to work carried out directly for Adedibu and remunerated by him (Animasaiwun 2013, pp. 12–15). Adedibu also supported a union NURTW's faction that was heavily involved in a series of murdered attacks and street wars between 2000 and 2011 (see below).

Even if Adedibu had no qualms about using violence, patronage and clientelistic networks cannot be built through coercion alone. Indeed, the absence of any reaction in Ibadan to the PDP victory in 2003, in spite of widespread electoral fraud, was said to attest to greater acceptance of patronage policies by the federal government under the Fourth Republic.[6] Clientelistic networks included a set of mutual social obligations, a dimension that some of the literature on neo-patrimonialism has neglected. In Ibadan, these social and moral obligations arise particularly between godfather and godson, in this case between Adedibu and the governor whose election he had ensured and their respective clients. Furthermore, he was challenged by other local figures who were competing with him to exert influence on governors such as Arisekola Alao, another influential big man (Roelofs 2016, p. 104).

Thus, in 2006, Adedibu withdrew his support from Rashidi Ladoja, the PDP governor in place, who was perceived as a greedy man who had chosen 'to eat alone'. More specifically, he criticised him for transferring only 10 million nairas (75,000 euros) a month to the budget of Oyo state instead of the 15 million nairas (100,000 euros) initially planned to reimburse the financing of the 2003 electoral campaign. Adedibu made no secret of his views in the national and foreign press.[7] No doubt his close connection to the President of the Republic protected him from being brought to trial, but his frankness also reflected the process of self-celebration rooted in the culture of *oriki*,[8] which extols the big man's capacity for redistribution and generosity (Barber 1991). As Karin Barber says, money is one of the best ways of gaining public acknowledgement as a big man, but 'having people' is that acknowledgement itself (Barber 1991, p. 183; Guyer 1995). Adedibu truly embodied the big man's radical transformations: 'having people' no longer means demonstrating valour in war or possessing magic powers as in the nineteenth century; it depends more on one's ability to build, maintain and expand networks of influence in institutions, local governments and city affairs, and political parties, and his ability to find work for people, resolve their problems with the police, and send their children to school (Barber 1991, pp. 240–243).

Diverting 0.2% of the budget of Oyo state (which amounted to 60 billion nairas in 2006) to redistribute it to tens of thousands of clients, friends, faithful followers and relatives is a form of patronage based on the embezzlement of public funds, but it appears far less illegitimate than the prebendalism of the governors whom these same people criticise for 'eating alone'.[9] Adedibu did not present himself as a member of a wealthy elite, obsessed with material accumulation. Demanding more of the governor and advertising it on the radio showed his determination to widen the circle of loyalties required to enhance his reputation and his power, based on public acknowledgement as well as on the transparency of financial redistribution to his clients for services rendered[10] (Figure 5.1). He was the incarnation of the patron figure in Yoruba society, chosen not only for his solid financial footing and his gift for building connections with the federal government, but also for his ability to give advice, play a model protective role (Omobowale and Olutayo 2007) and socialise the excluded (the unemployed and uneducated) around his person. He was a more traditional figure, a leader with no university degree, attached to the symbols of local Ibadan culture in his dress and speech (Roelofs 2016) who promoted 'sons of the soil' over educated foreigners (Chapter 6) and whose primary motivation could not be reduced to accumulating a personal fortune through the systematic use of violence.

Adedibu's death in 2008 signalled the start of a political recomposition. The PDP governor, Alao-Akala, no longer considered himself accountable to

FIGURE 5.1 Lamidi Adedibu rewarding his followers after the election of the governor in the state of Oyo, Molete, 22 April 2007 **Source:** Photo by Laurent Fourchard.

anyone, and his generosity, which once flowed through Adedibu's network, began to taper off. With the appointment of a new director of the Economic and Financial Crimes Commission (EFCC) in 2008, the other political patrons, Arisekola in particular, proved more reluctant to support a politician who was henceforth on the Commission's radar screen, and instead backed Abiola Ajimobi, an APC candidate who seemed less corrupt in the 2011 elections (Roelofs 2016, p. 109). That election signalled the provisional end of the PDP network's control over Oyo state. The 12-year record of the state's performance was extremely poor, implicitly pointing to the power of clientelistic rivalries and the strong hold of patronage and redistribution of federal revenue to networks of privileged clients. In 2005, Nigerian states were rated on their performance in the four areas of public policy: fiscal management, access to services, transparency and corruption; Oyo was one of the lowest-performing states in the country, on a par with the poorest states in the North (Roelofs 2016, p. 107).

The Metropolitan Project in Lagos

In 1999, Bola Tinubu, an international businessman trained in the United States and pro-democracy militant during the 1990s, was elected governor of Lagos state by winning the support of business circles and several organisations playing influential roles in local politics (market associations, road transport unions, OPC activists) (Nolte and Hoffmann 2013, pp. 12–13). He was succeeded by his godson Babatunde Fashola from 2007 to 2015. After decades of infrastructure crisis, these two governors both came to power offering a vision of metropolitan development and ushered in a new deal. Tinubu laid claim to the Awoist heritage of 'development' and 'civilisation' reflected in the term *olaju*, referring to the cultural baggage of the nineteenth-century missionaries and the expansion of mass education as a path to emancipation, an idea disseminated by Obafemi Awolowo's party starting in the 1940s (Adebanwi 2014; Peel 1978, pp. 146–147). The Lagos governors also incorporated the more recent discourse pertaining to competition among global cities, insisting on the need to develop modern infrastructures, access to basic services and planning agencies.

During his first term, Bola Tinubu sought to create a favourable environment for business and make Lagos state the benchmark of Nigerian progress 'through best practices, urbanism, and development' (Lagos State 2006). The rhetoric concerning competition among global cities has been more firmly embraced since then, which the state government has demonstrated by revitalising the old business district of Lagos Island and building a new one called 'Eko Atlantic' on reclaimed land as a centre for foreign investors and a gateway for Africa's emerging markets.[11] The two first governors of Lagos states had mega city ambitions: in 2015, the model to follow for Babatunde Fashola was London, New York or Dubai, rather than Delhi or Nairobi (Cheeseman and de Gramont 2017, p. 469).

The political break from the past called for by the governors relied on private sector growth, foreign investment and the state's ability to meet grassroots demands (Roelofs 2016, pp. 116–117). This break has been founded on reforming metropolitan governance and increasing the autonomy of Lagos state from federal allocations. Indeed, the portion of internal state revenues rose from 25% to 75% of the budget between 2000 and 2015, due to a considerable increase in taxes (from 600,000 nairas per month in 1999 to 15 billion in 2012). The increase of the personal income tax (from 500,000 to 4 million tax payers) has been the centrepiece of the tax reform (de Gramont 2015) followed from 2007 to a tax on property (land use charge), which has quickly become the second most important revenue stream for the state (Owen and Goodfellow 2018). Bola Tinubu substituted payment of taxes in cash for online payment into state bank accounts; he hired a private company (Alphabeta Consulting) to supervise the process, and replaced the existing administration with a new team managed by a banker (Cheeseman and de Gramont 2017). The transformation of the agency in charge of the budget – the Lagos State Inland Revenue Service (LIRS) – into a technocratic enclave isolated from the rest of the civil service and operating on merit-based hiring and promotion with high salaries was a key instrument in state tax reform (Adenugba and Ogechi 2013, p. 419). The agency's efficiency, attested by significant budget growth, did not necessarily imply a radical transformation of the state itself. It was rather a visible sign of what Erin McDonnell (2017) calls an interstitial bureaucracy, i.e. an administrative niche endowed with organisational characteristics grounded in a Weberian bureaucratic ethos (an impersonal, routine administration, effective and operational) within an institutional environment that did not necessarily resemble it. Tax reform served the state governing party's national ambitions as 'Lagos politicians were eager to demonstrate that they could deliver more benefits to their constituents than the PDP, reinforcing incentives to raise revenues for public services' (de Gramont 2015, p. 12). For Cheeseman and de Gramont (2017), there are specific conditions in which Lagos state governments were able to conduct ambitious reforms since 1999. The political stability of Lagos state government derived from the capacity of Tinubu and his network to gradually consolidate their hold over Lagos and encouraged 'governors to plan for the long term and to transform Lagos into a cutting-edge mega-city that they could be proud of, a drive that was underpinned by a distinctly high-modernist vision of the future'.

These policies did not eradicate patronage nor violence. Between 2000 and 2020, at least 150 000 people have been forcefully evicted from their homes in Lagos state in a brutal manner like during military periods with no warning, no resettlement and no compensation.[12] Lagos state did not make a seamless transition from a clientelistic form of government to an impersonal form of government based on efficiency, transparency, meritocracy and a bureaucratic ethos that transcended the long-standing ties of its civil servants. On the contrary, the period was characterised by institutional tinkering to combine a bourgeoning interstitial bureaucracy with traditional forms of patronage to serve the metropolitan project. This is especially visible with the creation and operation of 37 new local governments, urban renovation projects and the priority given to a new transport scheme.

The Constitution of the Fourth Republic recognises 20 local governments (LG) in Lagos state and the creation of any new governments requires an amendment to the Constitution. In 2004, Bola Tinubu decided to create 37 additional local governments. In response, the federal government decided to cancel the allocations intended for all local governments in Lagos. The Supreme Court in 2008 ordered the Lagos state government to eliminate the 37 new local governments that were instead renamed 'Local Council Development Areas' (LCDA) (Fourchard 2011b). The creation of local governments is a way of enlarging a party's political clientele on the metropolitan scale according to a model already tested during the Second Republic. They actually represented a significant financial expense. Since 2008, federal allocations for the 20 LGs have been transferred every month to Lagos state, notably to pay civil servants.[13] The LIRS divides the subsidies between the 20 LGs and the 37 LCDAs and a number of administrative personnel of the 20 LGs were transferred to the 37 LCDAs. Between 2004 and 2015, more than 22,000 civil servants and tens of thousands of contract workers were recruited in the low-income neighbourhoods of Lagos to work in LGs and LCDAs. Most of their wages are paid by taxes levied on salaried workers rather than by internal revenues generated from low-income neighbourhoods. According to Cheeseman and de Gramont, Tinubu had little to lose from increasing taxation of the formal sector, as his political backing came mainly from the dominant informal commercial sector in Lagos (Cheeseman and de Gramont 2017, p. 450). But taxes have also increased for more modest workers as suggested by the example of Bariga LCDA.

This LCDA is located in a densely populated, low-income neighbourhood of Lagos where 80% of the population is self-employed[14]. According to department heads, the actions of the LCDA are limited in number and scope: there are few infrastructure projects for want of investments and actions are limited to public health drills once a week or offering a few training programmes each year to craftspeople. In many ways, the Bariga LCDA is a machine for recruiting and ensuring the loyalty of clients to the party administration and its president: it is an archetype of the way local governments work in Nigeria. If internal revenues do not exceed 4% of total revenue, the finance department head also acknowledges that all taxes have increased since the economic crisis of 2015 due to the fall of oil price. In 2016, he systematically mapped all commercial establishments in the LCDA. Petty traders are subject to weekly pressure from LCDA tax collectors: often unable to pay online, they are subject to the discretionary power of street-level bureaucrats like before. Even if bringing the informal economy into the tax net has proven difficult (de Gramont 2015), this contribution is a significant (but unknown) transfer of resources from the 'informal economy' to state officials (Meagher 2018), in this case to the bureaucratic expansion of the state.

At the metropolitan level, priority was given to transport infrastructures and renovation projects. This resulted in a cleaner, orderly and embellished metropolis in which recovered public spaces were turned into gardens, parks and shopping areas, while many poor residents were evicted to leave the space for large-scale developments with uneven success: Eko Atlantic project, the Lagos flagship project is still, ten years after its launching, a classical urban nightmare fantasy with only a few constructions (Watson 2014). New public policies were implemented by state agencies created by merging existing institutions with their own budgets and substantial power in the key sectors of transport (Lagos Metropolitan Area Transport Authority – LAMATA), urban environment (Lagos Waste Management Authority – LAWMA), and the police (Task Force and Kick Against Indiscipline – KAI).[15] Like in the 37 new local governments, these agencies have enabled a number of unemployed workers and also delinquents known as 'area boys' to be integrated into the state apparatus (de Gramont 2015; Ismail 2009, p. 481).

In terms of public transport, the two first governors ensured the development of a new plan for the metropolis, which has been presented as a success story. It consisted mainly in creating regular bus lines that run in reserved lanes – Bus Rapid Transit (BRT) –, a plan inspired by the Trans-Mileno experiment in Bogota. In 2008, Fashola officially launched the BRT, a 22-kilometre bus line between Lagos Island and the mainland neighbourhoods. This was made possible because governors succeeded in maintaining the loyalty and fidelity of the principal road transport union leaders (NURTW) and integrating them into the new plan. The state transport agency, the LAMATA supplied the infrastructures (bus lanes, stops, terminals and repair shops) and the NURTW leaders agreed to the new regulations (bus lanes exclusively reserved during rush hours, queues at bus stops, advance ticket purchase) and to purchase and manage one of the two bus fleets operating on the BRT. The project demonstrates a strong political will but its expansion is limited: the only lane carried 200,000 passengers per day in 2012 (Basorun and Rotowa 2012, pp. 82–87) against 7–10 million road passengers (Salau 2015, p. 133). This weak percentage of public transport organised by the BRT (5%) should be compared with the fact that 85% of road transport in Lagos are assured by commercial vehicles (buses, taxis, motorcycles), which are under the authority of the NURTW. While the communication plan of Lagos state government focuses on the necessity to pay taxes online to avoid street-level bureaucrat corruption, the ten million daily commuters and 300 000 commercial vehicles in Lagos are left to the discretion of union leaders who are free to collect their revenues as they want.

Revenues, Violence and Politicisation in Motor Parks

Since the late 1980s, the NURTW referred to as 'the Union', has replaced local government authorities in the management of the motor parks. It should be noted that the 1970s represent for most cities in the country the definitive decline of former means of municipal transport (buses and trains) taken over by a network of private owners of mini buses locally called *danfo* followed by a new set of other commercial vehicles (motorcycles known as *okada* in the 1990s and tricycles referred to as *keke nappe* in the 2000s). This privatisation of public municipal transport is part of a larger historical trend in the continent with the reduction of public spending, the declining of salaries and the multiplication of parallel and rival unions and competing private operators (Rizzo 2011; Rubbers and Roy 2015). What makes the management of motor parks distinct in Nigeria is a mix of features from a systematic and continuous outsourcing of municipal revenues to the union to its open involvement in patronage politics and violence.

Motor parks became a central place of politicisation during the Second Republic (1979–1983) when their management was taken over by transporters unions. This process started in Lagos as the capital was the place of two concurrent powers: the federal government, the president Shehu Shagari, and his party, the NPN on the one hand, and the governor of Lagos state, Lateef Jakande and the United Party of Nigeria (UPN) on the other hand. NPN decided to enlist the support of members of a new union, the NURTW created a year before, in 1978 under the leadership of Adebayo Ogundare, known as Bayo Success who was giving the assignment of winning all the motor parks in Lagos over the UPN (Albert 2007, pp. 129–130). He did so in mobilising his large clientele of drivers during the 1979 electoral campaign and in resorting to violence in most motor parks in Lagos. NURTW was also able to extend his operations to five other south-western states (Ogun, Oyo, Bendel, Ondo and Kwara). Relationships between union leaders and political leaders have become closer since the return of a civilian regime in 1999, as the union is seen as the central player without which a governor's election cannot be won. State chairman of the union commands loyalty and obedience from members (Animasawum 2013), which makes the union attractive and strategic to those who aspire to become governor as a huge reserve of men can serve as political foot soldiers (Omobowale and Fayiga 2017).

Like vigilante groups seen in Part II, the Union could be seen as a 'twilight institution' (Lund 2006): it operates between state and society, what distinctly characterises it is its movement in and out of a capacity to exercise public authority. Over time however, the Union exercises an almost exclusive authority over the garages. This is the case in Lagos where the Union works hand in hand with the Lagos state government since 1999. In other states, officials sometimes temporarily suspend the activities of the union and their authority most of the time for security reasons. The governor of Oyo state Abiola Ajimobi proscribed the union in 2011 because it was involved in a series of fights that led to more than 70 deaths between August 2010 and June 2011. Oyo state transport department took over the control of motor parks until a new chairman was selected by the governor[16]. In April 2019, the new governor of Oyo state Oluwaseyi Makinde (PDP) again banned the union for the same reasons and replaced union members operating in motor parks by local government officials[17]. If faction fights are presented as serious threats to security, in both cases, the incumbent governor wants to get rid of former union chairman too close to the outgoing governor and appoint a new chairman loyal to his authority. In other words, NURTW with a few exceptional periods has been exercising an exclusive authority in Lagos and Ibadan motor parks despite having no constitutional power to do so. This grey zone between legal and illegal is what makes the union both powerful and contested. From drivers' and passengers' point of view, fees collected by union members are considered as extortion or theft (Agbiboa 2018, 2020). Union leaders and members claim instead to levy taxes 'on behalf of the government' and to be key players in organising transport. This fundamental ambivalence allows the union to present itself as an arm working on behalf of the state and the drivers while the popular understanding of union organisation remains associated with violence and extortion.

Extorting Money or Levying Taxes?

Union members act as unconstitutional tax collectors for state and local governments. According to the second and the fourth Republic constitutions, levying taxes in garage is the sole responsibility of local governments. As mentioned above, the control of motor parks was transferred to the union for political reasons in south-western states around 1979. This delegation of authority anticipated the recommendations made in 1985 by IMF to the federal government to reduce state and local government expenditure and staff. In the early 1980s, a few governmental staff were still operational in the motor parks of Owerri (a large city in South East Nigeria) even if a growing number of intermediaries were used to levy some fees (Okpara 1988). Since then state and local government authorities have outsourced the collection of a number of taxes to intermediaries, a common practice denounced by a recent Nigerian governor's forum report:

'In nearly all States, with potential tax payers consisting mostly of informal operators, governments try to minimize their own challenges by reducing interface with individuals. Actual revenue collection in many sectors is farmed out to agents and touts, sometimes as political settlement and patronage. Each 'authorized' group of agents simply print their own receipts for revenue collection, and molest potential taxpayers with a view to extracting maximum rent. As there are no strong checks, leakages remain high and collection efforts are concentrated on a few sectors where heavy investments are not needed before milking. These include stalls, markets and parks where potential taxpayers can easily be found and molested; and where extraction is easy because the taxpayer has high incentives to want to keep his business running'[18].

While this report rightly indicates that markets and motor parks are the easier targets for extracting money from business operators, the report however confusingly merges these too very different situations. In Lagos and Ibadan collecting taxes in most markets have not been transferred to private bodies but is done by local government officials and contribute to their internal generated revenues (Fouchard and Olukoju 2007). Revenues in motor parks are not collected by 'touts', 'informal operators' 'printing their own receipts' a view shared by the population, the media and part of the academic literature but by a bureaucratic organisation, the union, based on a set of illegal arrangements with elected politicians and central state institutions such as state, local governments and the police. Qualifying these relationships as informal is of little help and tends to obscure rather than clarify their very nature.

The everyday working of the union is based on two opposite cornerstones: contractual non-written arrangements between union leaders and government officials on the one hand, official tickets issued by the government and 'sold' in garages by union members on the other hand. The most systematic form of non-written arrangements happens to be at the local government level between branch chairmen and local government officials. Bunches of local government tickets are collected by union branch chairman in each local government on a regular basis (once or twice in a month). Once the tickets are sold out, the chairman gives the money back to the local government before collecting another set of tickets. The union takes out a commission (from 25% to 30%) from the official price ticket. This is the norm in all the parks visited in Lagos and Ibadan. In Lagos, local government tickets are sold 200 naira for a *danfo*, in Ibadan 100 naira for a *danfo* or 50 naira for an *okada*.[19] In sum, local governments illegally outsource to the union the power to levy taxes on their behalf.

This ambivalence is central in the politics of naming these unofficial tax collectors. There is a battle of legitimacy between the union and the public on that matter. In popular parlance, they are referred to as *agbero*, which means in Yoruba someone who helps or calls the passenger to enter the bus (*agbe* means to carry or to lift *ero* means passengers). The term still fits as *agbero* call passengers by announcing the direction of the route but *agbero*'s functions are essentially seen by the public as collecting money from drivers. The meaning of the term was probably changed with the transformation of *agbero*'s functions when they started collecting 'taxes' for the union in the context of the 1980s and 1990s economic crisis and increasing urban poverty. Their reputation as 'touts' has been reinforced by their violent behaviour of collecting money from recalcitrant drivers. But *agbero* insist they are not 'thieves' but 'notables' doing all it takes to survive in the motor-parks and getting a just reward for their labour (Agbiboa 2018). As mentioned by some executive members in one motor park in Bariga, 'the public see us as 'tout' but we are not. We are family men, we have wives, and our children are in school, we are living fine'[20]. Today no union leaders use the word *agbero*, which is too derogatory and much associated with violence. In Ibadan, 'tax collector' was the most common term used in 2018 when there was a peaceful cooperation between the governor and the union. This is not the case anymore. In 2020, with the banning of the union and the levy of taxes in motor parks by local government staff members, no union leader dared to qualify their members as 'tax collectors' anymore. In Lagos *agbero* are simply referred to as 'park attendants' an euphemism to deliberately side line their revenue functions known as being forbidden by politically aware union leaders. This politics of naming reveals the wish to distance the union to extortion and reflects a dual process of state appropriation (for tax collector) and depoliticisation (for park attendant).

The power to levy taxes is the legal basis against which a systemic illegal extortion is organised. Multiple expressions are used by *agberos* and drivers alike to describe the fees collected on the road: '*owo union*' (union money), '*owo* booking' (booking fee), '*owo* loading' (loading fee) '*owo* dropping' (dropping fee), '*owo* weekend' (money for weekend) '*owo* sanitation' (money for sanitation) '*owo* security' (money for security) (Agbiboa 2018, p. 72). What is fundamental is that the power to raise taxes for the government is not dissociated from the power to collect fees for the union. Once official taxes are collected, the real business of *agbero* start with an incredible level of discretionary power exerted on site by unit chairmen who fix the amount of what is referred to in Ibadan as 'not receipted fees' and in Lagos as 'dues' or everywhere in Yoruba as *kó èrò kí o sanwó* as 'load and pay'. Each time a bus loads passengers, the driver has to pay for the union. The amount depends of the type of vehicle, of the route, of the distance and of the transport fare paid by passengers. For intercity transports as well as urban buses, the union takes a percentage of the total sum of the transport fees for each departure. The most common fare in Lagos and Ibadan for a *danfo* of 14–18 passengers paying a 100-naira transport fare is the price of one or two passengers (100 or 200 nairas). A driver who stops ten times a day in a garage pays 10 times the 'due'[21]. For drivers loading passengers in a garage, the *agbero* marks the plate number of the vehicle on a small piece of paper. For drivers loading passengers on the road, the *agberos* have a distinct signature on the body of the vehicle or on its widescreen to recognise vehicles that have cleared their fees[22].

While union claims to levy taxes, in *agbero* parlance, however, points of revenue are referred to as *Oju iso* or 'sells points' or 'shops' in Yoruba. As a matter of fact, the money is not used to invest in the improvement of the park or to provide specific services for the drivers. The union has some social schemes for their members: loans to buy vehicles or insurance schemes to cover drivers in case of accidents. But these are meant for a very small minority. Drivers are not considered to be union members but have to first be recognised as 'loyal' by the chairman before joining the union. And among union members only the happy few connected to bigger patrons can benefit these services. In other words, money given by a driver is extortion because the fees are not related to any right or promise of service on the part of the driver (Varese 2014).

The amount of money made by the union at state, branch or even unit level is unknown as the union accountability is totally obscure. Very partial data produced by journalists shows a considerable amount generated daily by the union. According to one informant, NURTW Lagos state pockets up to five million naira daily from the different units under him (Agbiboa 2018, p. 76). This is probably a very low figure. Other calculations made by a journalist investigating in one of the largest garages in Lagos in June 2017 came to the conclusion that the total amount generated by the union was five million nairas a day for that one garage only.[23] If this is the case, we can extrapolate that the money generated daily by the motor parks in Lagos are closer to several hundred million nairas daily (several hundred thousand dollars a day).

An unknown part of this money is distributed to different state institutions, including to members of parliament, of the senate and to state governments, but this has to be further investigated.[24] Arrangements with the police is instead systematic. Each motor park gives a regular contribution to police officers passing by and to nearby police stations. The most common expressions 'security fees' or 'to settle the police' reflect illegal and widespread petty corruption practice. As suggested by local leaders:

'We assist the police to buy the fuel for their vehicles. This is understandable, they have their problems too. This also helps to keep good relationships with the police[25]. 'We have to be friends with them. We have to have a good rapport with them otherwise we will be having problems every day'.[26]

In Lagos most union members admit that they pay on a daily or weekly basis the NPF, LASTMA, the Task Force, the Vehicle Inspection Office (VIO), the famous Special Anti-Robbery Squad (SARS) and, in some occasions, the federal army. The daily amount is 1000–2000 nairas in each unit and for each police force in all the motor parks visited in Lagos. The amount given directly to the commissioner at the police station is unknown but union members confess that this is much more than the daily fees. The police extorts money from the union the same way union extorts money from the drivers: it is a racket with very uncertain outcomes. Most of the time it does not bring any service in exchange. Sometimes, 'security fees' provide protection against further police harassment as it helps to turn high official fines (5000–30 000 nairas) into a smaller bribe (2000–10 000), not receipted by police officers.[27] Arrangements between police officers and union leaders are everything but smooth but the system of outsourcing taxes to the union eventually bring large financial interests to many governmental and state authorities: regular and systematic bribes to the police, not recognised bribes to state executive and legislatives and tax illegally collected for the local governments.

The ways the union works in the public space help to rethink the practical arrangements between state actors and economic actors in the regulation of the economy. This seems to be very closed to the concept of 'informality from above', which, according to Ananya Roy (2009) is a form of deregulation, and a deliberate mode of state action on urban space, which materialises primarily through the state's fluid and arbitrary conceptualisation of legality and legitimacy. As Roy suggests qualifying planning in some Indians cities as failure is a rather short explanation as the state might pursue extralegal actions in its planning action and embodies a distinctive form of rationality (Roy 2009, p. 86). The problem is rather to turn a specific situational process into an ontological essence of the state as 'an informalized entity'. I suggest instead to follow Mathias Dewey (2018) and his call to explore sets of unwritten rules that are based on the suspension of the enforcement of the law that are directed towards specific social groups and enable behaviours otherwise regarded as illegal. La Salada market in Buenos Aires is the main supply centre for illegally produced garments (Dewey 2018). This market is illegal, as taxes and labour regulations are suspended to stimulate economic activities and increasing consumption of cheap clothes, which is supported by politicians in need of votes. Understanding unwritten arrangements to suspend the law helps to move beyond a normative vision reducing states in the South to weak states, weak institutions or institutions unable to implement the law. In the case of the motor park in Nigeria, the question is not so much that the law is not implemented (it could be implemented as it is the case in Ibadan in 2011 and 2019) but rather that it is suspended as it brings resources for the major actors involved in the arrangement. Local government officials outsource tax collections as they do not have the manpower to collect them but declare in their budget revenues illegally collected. These licit and illicit payments represent a significant transfer of resources from the 'informal economy' to state officials (Meagher 2018) and to top union leaders. No political and union actors have interest in stopping this transfer of illegal cash: it is undeclared and invisible unlike public federal money, which is increasingly controlled by the Economic and Financial Crime Commission (EFCC) even if this later remains weak, poorly funded and object of intense politicking (Enweremadu 2012). This blurring of the legal and the illegal is also what makes the union central in the government of transport.

Governing Transport Between Patronage and Bureaucracy

The money collected everyday has transformed the union into a key player in governing metropolitan transport and a powerful non-state bureaucratic organisation shaped by patronage and violence. The money is mainly used to pay an expanding bureaucratic organisation that mimics state apparatus in terms of structures, hierarchy, territory and vocabulary. It is hierarchical: units operating in a garage are controlled by branches that are under the authority of state chapters. Unlike the state administration where money circulates from the federal capital to the state and local governments, the money circulates from the street (the *agbero*) to the top of the office (state chairman). Each unit has to send a fixed amount of money collected in parks to the branch every week and each branch has to send an amount of money each week to the union secretary at the state level. Units and branches have revenue targets that depend on the number of vehicles registered and the profitability of the routes under their authority.

The patrons of the union at the unit, branch and state level is the executive committee (the Exco). An Exco is composed of a chairman, a secretary and a treasurer but most units and branches in Lagos or Ibadan include a vice-chairman, a financial secretary, a treasurer, an organising secretary, an auditor and often trustees or ex-officio members (often former chairmen). The number of staff in an Exco depends on the number of commercial vehicles registered in a garage. The average park unit has 30–50 registered vehicles and between 7 and 10 exco members. Most of the 40 members interviewed have been in the union for more than 20 years: they are all former drivers between 50 and 70 years old, with the rare exception that they are all males and most of them dropped out of schools. The union is a family that includes a larger network. Loyalty to branch and state chairmen and the capacity to meet the weekly targets over a long period of time are key to be promoted within the union. In the street, are *agberos* recruited and working under the authority of unit chairmen. Some of them have been doing this work for 20 years and many of them hope to be recognised and promoted as exco members. The number of *agberos* working in a garage depends on the size of the garage. In small units there are just one or two *agberos* in a larger unit, 12–15 of them. In Pako Aguda Unit or Odi Olowo Branch of Oshodi (Lagos) or in Sango minibus motor park (Ibadan), 3–7 *agberos* work from 6 a.m. to 8 p.m. one day on, one day off. Union members from *agbero* to chairman are paid by the 'dues' or 'non-receipted fees' according to local arrangements.

The union is not only using the state vocabulary to define the functions of its members, it has extended his reach following urban growth and the bureaucratisation of society. The multiplication of state and local governments during military regimes (1983–1998) was a way of popularising at the local level illegitimate military governments (Suberu 2001). Between 1985 and 1991, the federal government created 11 new states and 297 local government authorities across the federation. In Ibadan, the city council was turned into five urban local governments in 1991. The union soon followed the path: the single branch was split into five, one for each local government.

In the past two decades, branches for specific types of transport were also created (minibuses branches, taxis branches, intercity branches) 'to ease of management as there were too many buses and taxis in Ibadan'.[28] The creation of 37 new local governments in 2004 in Lagos turned into LCDA in 2008 has been replicated in Oyo state with the creation of 14 LCDAs in 2016 in addition to the 11 official local governments. The Union has again followed the path in collecting taxes on behalf of these non-recognised local governments as they need to increase their internal revenues in the absence of monthly federal allocations. In Ako Aguda unit in Lagos, drivers pay either for Surulere local government or for Itire Ikate LCDA or for both if their routes are under the authority of the two councils (Figures 5.2 and 5.3). In Ibadan, drivers pay to either Ibadan North local government or Oke Badan LCDA. Over the past 40 years, the expanding number of bureaucratic offices has been followed by

FIGURE 5.2 Surulere local government motor park tolls (2018).

FIGURE 5.3 Itire Ikate LCDA motor park tolls (2018).

the growing number of points of extortion in the street. In this process both the union and the state play out the boundaries between the legal and the illegal: the union collects taxes on behalf of local governments and non-constitutionally recognised LCDAs without making any distinction between the two.

The union governs the transport system in Lagos and Ibadan as it controls over 80% of road traffic and responds to the increasing need for mobility for a fast-growing urban population. Union leaders see the multiplication of branches and units as a way of extending their patronage networks throughout the city, to reinforce the rootedness of the union into society, to accommodate more members in the union and to collect more money. Each branch has an important level of autonomy in collecting fees and dues and opening new routes in areas poorly connected to garages. Local leaders sometimes provide 'fiscal exemption' as an incentive for drivers working on new routes: they don't pay the dues to the union before the route becomes profitable. Units and branches can henceforth welcome new drivers and follow the geographical expansion of urbanisation in a fluid and quick way. The union plays a no less important policing role among the drivers. The registration of motorcycle, tricycle and mini-bus drivers is compulsory. Union members collect the registration fees while the union driver data base is used by the police to trace criminals. They impose that their vehicles be painted with the official colour required by the state and organise the parking of vehicles, the waiting time, departure, routes and tickets fares. They solve conflicts among *danfo* drivers and between them and other commercial drivers: they don't allow – and sometimes take away motorcycles and tricycles – of those who try to pick passengers in front of their motor parks. They discipline drivers and try to convince them to have their vehicle particulars in order and to follow traffic regulations to avoid being harassed by various security forces. Motor parks governed by the union appear to be the most efficient street-based bureaucratic office: the 16-hour presence of exco members (from 6 a.m. to 8 p.m.) allows implementing a daily routine done without papers mainly based on their personal knowledge with the drivers (Figures 5.4 and 5.5 in Oshodi).

Apparently, NURTW works like many other associations of drivers in the continent. But unlike in Senegal where the local administration partly managed garages, in Lagos and Ibadan, union members have a total autonomy. They are the exclusive actors with direct contact with drivers and passengers; they act as liaison agents between union chairmen, government officials, drivers and passengers. NURTW is a twilight institution and union members act as street-level bureaucrats appropriating state vocabulary, hierarchy and codes. But they are bureaucrats of a special type. They work full day for the bureaucracy of NURTW and invest a very limited time for the government; their loyalty goes to the chairman of the park not to local government civil

FIGURE 5.4 Odi Olowo motor park (Lagos) **Source:** Photo by Laurent Fourchard.

servants; they implement union decisions taken after unclear and uncertain arrangements between top officials and top union leaders, which are unknown to the public. They follow the specific and peculiar private agenda of NURTW union leaders, which are conditioned by the will to control motor park revenues in using their own political relationships.

Violence, Loyalty and Politicisation in Motor Parks

Money and violence are intimately linked since the inception of the union in 1978. Collective violence is mainly linked to fights between union members. Large-scale violence operates in Lagos and Ibadan as state capitals are the core of urban taxation and of harsh competition between concurrent networks. Violence could occur during election campaigns but on a more regular and

FIGURE 5.5 Oshodi Isolo Motor Park (Lagos) **Source:** Photo by Laurent Fourchard.

larger scale in situations where there is a change of political party at the state level. Ibadan is a typical case in Nigeria as violence has been systematic between different factions that are themselves supported by one big patron in the state (a governor, a candidate for governorship, a godfather) and their political parties in power (AD: 1999–2003, PDP: 2003–2011, ACN: 2011–2019). This last section does not review the forms of violence happening in motor parks and streets but tries to understand the ways in which controlling revenues has become part of larger political conflicts especially in the divided city of Ibadan and to explore the importance of political loyalty in using the case of the current NURTW Lagos state chairman, a protégé of Lagos governors.

According to one of the organising secretaries in Oyo state, 'at the inception of the 4th Republic, NURTW teamed with the new AD governor in a relationship that saw the preservation of the leadership of the Union chairman in exchange for protection of members of the AD at political gatherings'. The first governor of Oyo state, Lamidi Adesina effectively got elected in April 1999 with the support of NURTW vice-chairman, Alhaji Lateef Akinsola alias 'Tokyo' (Albert 2007, p. 141). Adesina and Tokyo were political opponents during the military regime of Sani Abacha (1994–1998) and got to know each other in the same prison cell. Before the gubernatorial election of 1999, Tokyo and Adesina passed an alliance to overthrow James Ojewumi, the Oyo state union chairman since 1988.[29] When Adesina became governor, James Ojewumi, was literally 'forced off' from office on 17 August 1999 by Tokyo thugs who, thereafter, systematically resorted to violence to eliminate Ojewumi's network in the city. On 26 June 2000, Tokyo thugs disrupted a meeting where Ojewumi loyalist members were set on fire, their vehicles burnt and their office destroyed. Despite the fact that 40 Tokyo members were arrested and taken to jail, Tokyo was never arrested and instead stayed as NURTW Oyo state chairman during Adesina's mandate. The lack of justice and the impunity against perpetrators of criminal acts is common in Nigeria. In this case, it is the very product of patron–client relationships in which the protégé of the governor had no risk of being arrested. This impunity had dramatic consequences: the politicisation of motor parks authorised the control of its revenues by all means and triggered non-ending street wars between NURTW factions.

In this context of nascent rivalry between two existing factions, Adedibu aimed in 2001 while contesting from the presidential party (PDP) against the regional party (AD) to win Oyo state. He supported Ojewumi's faction in an attempted coup to overthrow Tokyo aligned AD faction and to uproot his members for the control of parks in October 2001. It failed and left behind an incredible number of deaths: between several dozen and three hundred (Albert 2007, p. 142). Similarly, the 2003 elections were held to be largely rigged like those in 2007, thanks to Adedibu's support to the 'Union PDP' faction. The rivalries did not stop after Adedibu's death in 2008, as NURTW was now divided into three factions supported by different governors. The Nigeria Watch database, which tracked the number of violent deaths based on Nigerian newspapers, reported 21 clashes between factions of the NURTW in the state of Oyo between 2007 and 2011, causing the death of 100 people. It is only when governor Ajimobi banned the union for some months and promoted a more consensual chairman that a relative peace came back to the city between 2011 and 2019. In other words, the union in Oyo state has become a central key player in election campaigns while patronage and violence have *de facto* politicised factions' members. The patronising of the union in Ibadan looks like Oyo state government action: it lacks a long-term project.

NURTW in Lagos is also considered as a major supplier of thugs recruited among its members to advance the Lagos state governor's electoral campaigns (Agbiboa 2018). But unlike in Ibadan, union leaders have developed a strong relationship with the dominant political party (AD/AC/ACN). Any attempt by the PDP and the federal government to uproot the union from the influence of the regional party has failed. The killing of Saka Kaula in January 2008, Lagos state chairman and AC supporter, led the following week to clashes between factions supported by the two rival parties and to the postponement of local government elections by the state governor.[30] Unlike in Ibadan, however, party-based violence are the exception in Lagos due to the incapacity of the PDP to find proper support in a number of garages. While the history of Ibadan shows strong rivalry along partisan lines, NURTW in Lagos is a history of monopolisation of motor park revenues by a faction led by the most popular union leader in 2020: Musuliu Akinsanya.

Alhaji Musiliu Akinsanya, alias Olu Omo, was an area boy at the Oshodi market when he joined and worked for Bayo Success faction between 1978 and 1983 until he was arrested by the military government in 1983. In 1996, he became the leader of Oshodi One Organisation, a vigilante organisation that consisted of a group of 500 former area boys, and paid by the traders to ensure the security of the market at night, a task it performed for ten years by developing a relationship with the Oshodi police and the main market representatives (Berg et al. 2013, p. 175). He became Chairman of one union branches in Oshodi at an unknown date. In early 2008, following the assassination of the state chairman, Musiliu rose to the position of Lagos state treasurer. The best position within the union is to be able to be in a key position in one big motor park and in office: Musiliu has cumulated the money associated with these two functions as branch chairmen of Oshodi large garage and as treasurer for the state chapter. He was already a well-established union leader when Louis Theroux, a BBC journalist met him that year: he headed a network of thousands of *agberos* and *exco* members and maintained a network of relations with market women, area boys and imams (Theroux 2011).

In 2017 he confessed to the press: 'Tinubu played a significant role in my life, my success and achievements. The governor is my father, my mentor and my role model'[31]. Several union members recognised that Mc Olu Omo has been selected as the state chairman in October 2019 because of his loyalty to Tinubu. Akinsanya was never arrested by the police despite the regular press reports against his men's involvement in a series of deadly clashes: in Oshodi (May 2007, 5 death) or Lagos Island (January and February 2012, 15 death). After he was elected chairman in October 2019, he was given the support of the police to arrest branch members who were recalcitrant to his authority[32]. Musiliu, as an *agbero*-turned-state chairman epitomises the ideal career for thousands of union members: from the street to the position of a big patron in the state. Most of them see him as defending their precarious jobs in being close to the state government. He could easily be seen as an embodiment of the 'discharge' of the state (Hibou 1999) or a political entrepreneur reflecting the fractured nature of state power in a contemporary world metropolis (Weinstein 2014).

Local union leaders know that NURTW is an institution with important bargaining power. The union was a key player in the making of the BRT in Lagos. The opening of BRT bus stops closed to important motor parks in Obalande, Oshodi, Oyingo has reduced both the traffic of *danfo* from 30% to 50% and the money collected by exco in bus terminals in the last five years[33]. While local members are open to discuss illegal arrangements with local officials and criticise police extortion, they are not vocal against the BRT nor against the government. They claim instead that the union and the state government is one. This lack of criticism should be seen as part of a deal in which loyalty is central in the working of the union. Union members interviewed in the motor parks of Lagos after Musiliu's election declared: 'he is our governor', 'he is our state chairman; we must follow his order'. 'This is where we work, we must follow him'. 'We have his badge, we have a banner to show our support and loyalty'.[34] Everybody knows in the motor parks the close relationships between Olu Omo and Tinubu the two major patrons of the state. Most of union members want to show their loyalty, as is the state of poor and rich union members to be included and remain in a powerful organisation that gives to hundreds of thousands of members a relative job security in a radical world of street life and uncertainty.

Conclusion

The regular payment of federal allocations and the growing wealth of metropolitan revenues guarantee Lagos a high degree of financial autonomy and allow the governors to implement a project that marks a rupture from earlier periods in terms of urban planning, the provision of infrastructures and services and the recovery of public spaces. The state is reasserting its capacity to intervene without abandoning the coercive practices of certain agents especially in forced evictions. To implement their project, the governors agreed to the inclusion of many groups of influential actors (transport unions, market associations, LCDAs, members of the OPC) in their political networks. The case of Lagos seems to indicate that wider redistribution policies have partly replaced the patrimonial policies reserved for a few privileged clients during the military dictatorships. In Ibadan, the return to democracy has not fundamentally altered the governors' prebendal practices, even if it has probably accentuated the redistribution of federal revenue through Adedibu's clientelistic networks. His patronage has facilitated this redistribution in some neighbourhoods and politicised parts of urban spaces, while intensifying factionalism and violence in the transport sphere to control taxation.

The example of Lagos demonstrates that patronage and clientelism are not only compatible with urban development, but actually support it. Sources of change in state working have come from the transformation of institutions (strong bureaucratic growth, the development of bureaucratic niches), the implementation of public policies, the development of a megacity agenda by Tinubu and Fashola and from an expanding bureaucratic machine (introduction of new local governments and new state agencies), which have absorbed former street urchins (area boys, delinquents) and local leaders supposed to be loyal to the governors' party. The metropolitan project has reconfigured the state (by creating several ministries and government instruments) (King and Le Galès 2011, pp. 468–471), and accounts for the population's relatively new consent to pay taxes chiefly for services.

This is one side of the Lagos success story coming mainly from policy and officials' circles. Behind the BRT flagship, the government of transport has been left to the discretion of the Union that has a free hand to raise revenues from the daily transport of millions of Lagosians. For a part, the union is a source of social and political inclusion. It is not radically different to LCDA or some state agencies like KAI, known for being a source of employment for tens of thousands of illiterate young male adults who otherwise would be roaming in the street. To be included in the union is being part of a family, it is a way to become somebody and after a life of driving to stabilise revenues in a very precarious urban environment.

The union is also a source of systematic illegal and violent practices allowed or supported by the elite in power. If the union is like a state bureaucracy and union members act as street-level bureaucrats, their daily work based in motor parks and the streets is fundamentally based on a suspension of the law, which is of benefit to different official networks. An increasing number of studies show that tax compliance is high in Lagos. But there is also a consensus of the millions of passengers and drivers to consider 'dues' as illegal extortion, a known practice by state governors and administration in entire contradiction with the state public campaign inviting Lagosians to willingly pay their taxes. 'Dues' and the money made from the streets are meant to pay an expanding bureaucracy and to enrich the most powerful men in the union, most of them having left the hard life of motor parks. After Olu Omo won his election as chairman in October 2019, a number of journalists criticised him for having accumulated immense wealth, enough to pay tuition for six of his nine sons and daughters in American universities. Like Tokyo during the Adesina mandate in Ibadan, Olu Omo has never been charged before court despite the fact that a number of murders involving 'his boys' have been reported by the press. Motor parks are *de facto* places of an exclusive NURTW authority in which violence against opposed factions or drivers reluctant to pay their 'dues' is never sanctioned when used by factions loyal to the governor. The power of the union is not only based on its capacity to use violence in public space, but in being located at the crossroads of the most powerful patronage networks and very formal institutions in the city. In raising taxes on behalf of local governments, in giving back part of the 'dues' to police and state executives, in being loyal to governors according to arrangements who gave them the public authority to run motor parks or in negotiating with the governments the implementation of the BRT, the union is a central player that no political or bureaucratic power is willing to get rid of.

Notes

1. With the exception of an increase of 13% of federal revenue from petroleum and reallocated to the states of the Niger delta.

2. The distribution under Abacha was as follows: 48.5% (federal), 24% (states), 20% (local governments), 7.5% (special funds). Since 1999: 48.5% (federal), 26.7% (states), 20.6% (local governments), 4.1% (special funds) (Lukpata, 2013, p. 32–38). Since 1996, the federal government also collects a value-added tax (VAT), keeping 15% and allocating 50% and 35% to the states and local governments, respectively (Suberu 2013, p. 93).

3. For the state of Lagos, 48 billion nairas in 2000, 112 billion in 2005 and 662 billion in 2016. For the state of Oyo, 39 billion in 2005 and 165 billion in 2016.

4. The Dawa or 'call' in Arabic designates a proselytising technique used by various Muslim movements to broaden their scope of dissemination.

5. Ruth Watson notes that there was always something to eat at the house of a warlord for his thousands of followers in the early twentieth century (Watson 2003, p. 477).

6. The 2003 elections were held to be largely rigged, like those in 1965–1966 and in 1983, except that unlike the earlier ones that were followed by violent and massive protests that ended the First and then the Second Republic, those in 2003 were not followed by demonstrations in Ibadan or in the other South-West States that had gone over to the PDP (Nolte and Hoffmann 2013).

7. Observation at the residence of Adedibu in Molete, April 2007.

8. *Oriki* is a genre of Yoruba oral literature generally used in prayers of praise for individuals or communities invoking their bravura, origin, reputation or success.

9. Interview with Adedibu supporters, April 2007.

10. Observation at the residence of Adedibu in Molete, April 2007.

11. http://www.ekoatlantic.com/about-us/

12. Justice and Empowerment Initiative, https://www.justempower.org. I thank Elsa Rousset, urban planning fellow, for sharing information on the scale of forced eviction in Lagos.

13. Interview with the clerk officer of the LCDA in Bariga, 4 December 2014.

14. The census carried out in 2006 by the statistics department of the State of Lagos en 2006 claimed the figure of 1.02 million inhabitants for the local government of Somolu/Bariga, out of a total metropolitan population of 17.5 million inhabitants. It is one of the five most densely populated neighbourhoods in Lagos. 80% of the adults are self-employed.

15. The Task Force (or Lagos State Environmental and Special Offences Enforcement Unit) is in charge of enforcing environmental laws and curbing special offences in the State of Lagos, which include street trading, illegal occupation of public space, annexation of public space, land extortion, waste disposal, public prostitution and illegal street closure. It also oversees slum demolition. The Kick Against Indiscipline Brigade (KAI) is a unit of the State of Lagos Ministry of the Environment specifically responsible for eradicating street trading, demolishing illegal constructions on pavements and eliminating prohibited waste disposal. The duties of the two bodies overlap to a great extent.

16. Interview with assistant administrative secretary, OYO state Chapter, Ibadan, January 2018.

17. Interview with local government officials and several NURTW branch chairman, January 2020.

18. Nigeria's Governor Forum. Internally Generated Revenue of Nigerian States –Trends, Challenges and Options, October 2015, 18 March 2016, p. 21.

19. Interview with Branch Chairman Akinyele Minibus in Akinyele NURTW's office, Ibadan, January 2018

20. Interview with Exco in Bariga, Lagos, January 2020.

21. Interview with Exco of the Pako Aguda Unit, Surulere, Lagos, January 2018.

22. One of the best illustrations of this practice by *agbero* is shown in Alain Kassanda's excellent movie 'Trouble Sleep', 2020.

23. 'Ban on NURTW in Oshodi costs N5m daily, affects 500 jobs', *Nigerian Tribune*, 10 June 2017.

24. Interview with two chief advisors of Oyo state ministers. January 2020.

25. Interview with Chairman Branch of Sango Intercity Motor Park, Sango. January 2018.

26. Interview with Exco members in Bariga Unit, January 2020.

27. This is an argument systematically advanced in all the motor parks.

28. Interview with Chairman Branch of Sango Intercity Motor Park, Sango, Ibadan, January 2018.

29. Interview with Chairman Branch of Sango Intercity Motor Park, Sango, Ibadan, January 2018.

30. 'Lagos paralysed: NURTW chairman, Saula, murdered', *Nigerian Tribune*, 8 January 2008; 'Violence as thugs set Oshodi market ablaze – Over murder of NURTW chairman', *Nigerian Tribune*, 9 January 2008.

31. Musiliu Akinsanya, 15th May 2017.

32. Damilare Famuyiwa, 'How MC Oluomo made police to arrest 48 NURTW members', *Pulse Nigeria*, /1st September 2020.

33. Interview with exco members in Obalande, Oshodi, Oyingbo. January 2020.

34. Interview with Bariga exco, January 2020.

CHAPTER 6

Bureaucrats, *Indigenes* and a New Urban Politics of Exclusion

For the past four decades, Nigeria has engaged in the peculiar practice of dividing its citizenry into two categories: *indigenes* and *non-indigenes*. The *indigenes* of a specific place are those who can trace their ethnic and genealogical roots back to a community of people that originated in that place (HRW 2006, p. 1). Everyone else is a *non-indigene*. The NGO Human Rights Watch (HWR), one of the first international organisations to express concern about the consequences of such a distinction between citizens, has shown that these policies have led to marginalising and excluding *non-indigenes* from access to employment, basic services and political contests and relegated them to the status of second-class citizens. This distinction between *indigenes* and *non-indigenes* has been institutionalised: the vast majority of the country's 774 local governments produce certificates of indigene or origin, which authenticate the holder's ancestry and guarantee him or her a number of rights. The certificates issued by local governments are at the heart of Nigeria's bureaucratic machinery because they are indispensable to meet its policy of hiring quotas in federal administrations, the army, or the police. They are equally required in state administrations and local governments, and to register at state and federal universities, in federal high schools, and if the HRW report is to be believed, they constitute the legal basis for excluding *non-indigenes* from state services and many other public goods.

In this final chapter, I will examine conflict and negotiation required to procure this document, what such transactions tell us about the forms of inclusion and exclusion that are approved by the local administration, and the extent to which this practice is perceived as discriminatory by *non-indigene* also referred as settler populations. I will observe this process from the inside, i.e. in local government offices where the certificates are issued in Ibadan, capital of Oyo state where this process is poorly contested and in Jos, capital

Classify, Exclude, Police: Urban Lives in South Africa and Nigeria, English Language First Edition. Laurent Fourchard.
© 2021 John Wiley & Sons Ltd. Published 2021 by John Wiley & Sons Ltd.

of Plateau state, where conflict over indigeneity is at the core of collective violence since 2001. Comparing the politics of classification and exclusion in Ibadan and Jos shows two radically different trajectories of belonging and competing definitions of urban ancestry.

In Nigeria, relationships between *indigenes* and *non-indigenes* are determined by controversies over how to determine who is a genuine *indigene* of the place (HRW 2006, pp. 43–44 and 54–58) and who can 'claim over ownership', two expressions that need clarifications. Even though the terms *indigenes* and *non-indigenes* are commonly used in Nigeria today, the history of these words does not exist and has no links with the French word *indigène* (which meant natives during the colonial period). I suggest instead that the categories of natives and non-natives so widespread during the colonial period in Nigeria (see Chapter 1) were probably replaced in the 1970s by the designations *indigene* and *non-indigene* due to three events: the decree of indigenisation promulgated in 1972, which gave Nigerians the exclusive right to ownership of certain companies in the country, the promotion of indigenous peoples at the United Nations during that decade and the integration of the term *indigene* in the constitutional debates that took place in the second half of the 1970s (see the following text). The expression 'claim over ownership' indicates a threefold claim of belonging to a place (a community, a city, a local government or a wider political area), ownership of that place (and therefore being in a position of authority to decide what is authorised and prohibited), and having one's ancestral origins in that place (and thus land rights as well as honorary and political rights).

Groups can contest the legitimacy of the government to monopolise the definition of the origin of people but looking at the very definition of who is an *indigene* from within the office has surprisingly received little attention. It is however a highly sensitive political issue in Nigeria as keeping resources for *indigenes* is both a strategy implemented by political leaders to exclude outsiders and a daily practice of bureaucrats who have the power to define who is an *indigene* and who is not. Each certificate of *indigene* is the embodiment of specific local ideologies, and reflects uneven forms of exclusion and inclusion of citizens in the local political community. On what grounds are these certificates delivered? What proofs of ancestry are used in the cosmopolitan urban environment of Jos and Ibadan? Who is considered an *indigene* of the city and to what extent this qualification is important in accessing resources in these cities?

Most of the literature has understandably insisted on the conflicts triggered by indigene–settler conflicts (Adebanwi 2009; Ekeh 2007; Ostien 2009; Ukiwo 2006). Excluding *non-indigenes* and considering them as second-class citizens led to repeated violence and sometimes radically changed the political landscape and increased city divisions especially in Jos (Higazi 2007, 2011; Krause 2016; Madueke 2017, 2018, 2019; Mustapha et al. 2018). Less has been written on the very definition of indigeneity at the local government or at the urban level, on the mundane practice of producing documents that discriminate millions of Nigerians and how *non-indigene* citizens in Nigeria challenge, negotiate or sidestep such an ordinary discrimination (Ehrhardt 2017; Fourchard 2015, 2018). The point here is not to minimise the power of discriminatory practices based on origin, but rather to bring out the fact that such discourse and practices do not necessarily lead to violence and conflict. It is actually hard to know whether discrimination affects the living conditions of *non-indigenes* regarding access to employment, university, political positions, housing, property and basic services.

This chapter aims to link the detailed topic of producing documents from within the office to wider politics of exclusion and bureaucratic power. The growing bureaucratisation of Nigerian society, made visible in the preceding chapters, is observed here through mass production of certificates and appears to go hand in hand with increasing capacity to classify citizens. In extrapolating partial figures, there are probably today more certificates of origin produced yearly by local governments (3 to 4 million) than birth certificates (around 1.5 million) (Fourchard 2018, pp. 77–79). This approach falls within the history of bureaucratisation experienced by ordinary people. According to Alf Lüdkte, the routines and rituals of state practice that Max Weber identified as the core of domination in everyday life played the precise role of making social boundaries, either those that are specific to bureaucracies or those determined by processes exterior to them, while they appear natural have already transcended (Lüdkte 2015a, p. 49). Lüdkte showed very clearly through the notion of *eigesinn* how individual appropriation of the structures of domination – namely bureaucratic structures – helps to legitimise those structures. In this case, the bureaucracy plays a decisive role due to its ability to take a constellation of contradictory interests into account. Local administration in Nigeria is considered inefficient and corrupt. 'You can buy anything in Nigeria, including certificates of *indigene*', suggests one politician in Ibadan, which sums up the commonsense attitude towards local administrations and is consistent with the conclusions of an HRW report on local ramifications of corruption in Nigeria (HRW 2007). Then why should we bother taking a closer look? Precisely because the daily transactions between bureaucrats and users cannot be reduced to acts of corruption (Blundo and de Sardan 2007; Chalfin 2010; Fourchard 2018). Instead, it poses questions on the power of bureaucrats and their degree of discretion on how to enforce rules and laws (Bierschenk and Olivier de Sardan 2014; Lipsky 1980) while opening at the same time the question of an 'ordinary citizenship in action' (Carrel and Neveu 2014).

Actually, anthropologists and historians have looked through paperwork but seldom look at it (Hull 2012b, p. 250; Kafka 2009, p. 341). Moving from high politics and broad institutional relationships to the humble stories of documents allows Matthew Hull to explore both crisis and stability in Islamabad (Hull 2012a, p. 4). Bureaucratic writing is commonly seen as a mechanism of state control over people, places, processes and things but the political function of documents is actually much more ambiguous. Instead of characterising documents as the passive instruments of bureaucratic organisations formed through norms and rules, Hull considers them as constitutive of bureaucratic activities and of forms of sociality (Hull 2012a p. 19). Every kind of graphic artefact has its own politics (Hull 2012a, p. 12), which has been hardly examined in African contexts (see however Awenengo Dalberto and Banégas 2018; Breckenridge and Szreter 2012; Cooper 2012).

There is behind the current politics of exclusion of *non-indigenes* a very unclear legacy of the past: in both cities, competing ancestral claims are rooted in the history of the city and colonial planning has reinforced mutual exclusivity between migrants and natives (Chapter 1). This legacy plays a more important role in shaping today's divisions than what has been said in the literature. It is not, however, a burden that contemporary society cannot get rid of but rather a register of the past that could be mobilised if necessary. Coming back to different historical layers in a comparative approach helps to see the ways in which history has been used in more recent times – since the 1990s – as a political and bureaucratic tool to exclude *non-indigenes* to access limited state resources. To explore this new urban politics of exclusion (different from the exclusion of migrants during the colonial period seen in Chapter 1), my aim is first to replace the production of certificates of *indigene* into the larger politics of indigeneity, second to understand how the politics in the office issuing certificates are determined by a more recent and politicised urban ancestral claims and third how the spatial urban divisions might be affected – or not – by this form of official discrimination. A systematic comparative approach helps to highlight radical different trajectories between Ibadan and Jos.

I have collected more than 40 interviews with local elected officials, local government heads, civil servants and chiefs in charge of issuing the certificates, observation of hearings granted to candidates with a view to obtaining a certificate in three different places in Ibadan North local government, one of the five urban local governments of Ibadan; in Akinyele, one of the 11 local governments of the metropolis of Ibadan at its northern periphery,[1] and in Jos North, the most contested urban local government of the capital of Plateau state. I also did additional interviews in neighbouring rural local governments of Plateau state (Bassa and Jos East). I have moved outside the office to try to understand the effects of this official discrimination on the everyday life of residents qualified as *non-indigenes*. In Ibadan, around 30 interviews were collected among political leaders, merchant associations and residents in two neighbourhoods inhabited mainly by populations of Hausa origin to explore their possible exclusion from political and administrative offices as well as for understanding their history. Sabo (in Ibadan North) is a neighbourhood that was designed by the colonial administration in 1912 to exclusively accommodate the Hausa population. Sasa (in Akinyele) is a neighbourhood that welcomed Yoruba farmers and Hausa traders in the 1970s. In Jos, I rely on second-hand literature to explore the effects of indigeneity politics on the transformation of urban space.

This last chapter refers to the historical institutionalisation of exclusion of the *non-indigene* during the colonial and immediate postcolonial periods before providing a detailed analysis of the contested politics of new urban belonging in the past two decades. It then shows how the politics of graphic artefact of certificates produced in local government offices are radically contested in Jos and poorly challenged in Ibadan. The last section highlights the opposite trajectory of indigeneity politics on the production of urban space. In Jos, two decades of collective violence have increased the spatial division while in Ibadan, both patronage and government action in access to urban services downplays the discriminatory exclusion of the *non-indigene* indicating the inclusive dimension of patronage mentioned in Chapter 5.

Institutionalising Exclusion, Manufacturing New Urban Belonging

While the word 'xenophobia' often refers to discriminatory discourse and practices with regard to foreign nationals, in many African contexts, analyses have been dominated more by the notion of autochthony. In contrast to xenophobia, autochthony expresses the claim of being the first to settle in a certain place and to have ancestral roots there. Very often autochthonous discourse appeals to a community imaginary shaped by a national and/or local historical narrative that constructs allochthones as a threat to the community and uses a variety of metaphors associated with order, social control, purity and public health to justify excluding them from the social body (Geschiere and Nyamnjoh 2000). Xenophobia and autochthony often rely on similar repertoires (ethnicity, territory, nationalism, ancestral ownership, etc.) that continually redefine the contours of citizenship. Nigeria's distinctions among African countries lies in having institutionalised the differences between *indigenes* and *non-indigenes* as forms of belonging that condition access to federal and local resources (Fourchard and Segatti 2015, p. 8). At the national level, indigeneity in Nigeria is effectively the product of a long history of indirect rule during the colonial period and the practice of regionalism in the 1950s and the 1960s (see Chapter 1). This legacy was transformed more recently by what is referred to as the Federal Character of the Constitutions of the Second Republic (1979) and the Fourth Republic (1999).

The Constitution of 1999 guarantees the citizens of Nigeria freedom from discrimination based on their ethnic affiliation, origin, sex or their religious or political opinions; it also guarantees that citizens may not be granted privileges or advantages because of their ethnic affiliation, origin, sex or their religious or political opinions.[2] In reality, the Constitution stipulates that administrative and political leadership positions in the federal government must be divided according to the origin of the individuals. This particular provision is related to its federal character, a notion invented by Nigerian constitutionalists in the mid-1970s and inscribed in the constitutions of 1979 and 1999. It aims to promote national unity and combat regionalism and ethnicity, which are considered to have been key factors in the collapse of the First Republic and in triggering the civil war (Kirk Greene 1983). A quota system was put in place in all federal administrations starting in the 1980s, and guarantees today the representation of citizens from the 36 states and the 774 local governments of the country. The Commission in charge of controlling the federal character ensures that no domination of one or more ethnic groups can take place within the federal government or its institutions (Mustapha 2007) even if the Federal Character Commission lacks funds, staff and political support to redress the existing imbalance in public service employment in states and local governments (Demarest, Langer and Ukiwo 2020). The application of this federal character implies defining what 'belonging to' a state and being 'indigene' of a state means. To date, there is no legal definition of an indigene person. This task has been left to the discretion of the states and local administrations.

Belonging to a local government area or a state allows an individual to pursue a political career or find employment in the local governments or the state. The fact of having only indigene persons working in the local administrations and the states is widely considered to be standard practice by civil servants and politicians in Ibadan and Jos. The history of the exclusion of non-indigenes from administrative positions is in fact older than the history of quotas in the federal bureaucracy. It is related to the fact that the 'sons of the soil' have always been favoured in gaining access to administrative positions, first in the administration of the Native Authority and later in local governments when they were set up (in 1953–1955 in the south-west including in Jos). The states (and before them, the regions) have had a tendency since the 1950s to favour the citizens of their territory by excluding citizens from other regions and states from access to services and positions in the local administration (Chap 1). It appears that this preferential treatment has been reinforced since the return to civilian government in 1999 as a method for reducing the number of candidates for political responsibilities and administrative positions (Ehrhardt 2014, p. 5; Fourchard 2018). The link between democratisation and the exacerbation of policies of belonging is significant here in Nigeria as in many other African countries where the use of the label 'foreigner' to disqualify or eliminate an adversary resurfaced following the liberalisation of authoritarian regimes (Bayart, Geschiere and Nyamnjoh 2001; Geschiere 2009). This overview is however too general and urban and local variations play a more central role in shaping the boundaries between *indigenes* and *non-indigenes*. Ibadan and Jos have a common colonial background but around the 1980s and 1990s new forms of urban belonging became dominant and led in Jos to one of the most tragic mass violence in the recent modern history of the country.

The top priority of colonial administration in the early twentieth century was to determine what should be done with migrants who did not come under the jurisdiction of native authorities. The local administration decided to grant certain migrants a separate place of residence (reserved sections of the city called Sabon Gari or Sabo) with specific institutions and rights that differed from those of the local population, thereby creating a lasting distinction based on people's native or non-native origin (Chapter 1). In Ibadan, people of non-native origin especially Hausa population were grouped together, in Sabo, which maintained its ethnic exclusivity throughout the colonial period (Cohen 1969). While Ibadan society was welcoming 'settlers' in the nineteenth century (Falola 1985), hostility towards 'strangers' developed during the colonial period largely supported by the native authorities (Adeboye 2003, p. 315). In the 1950s and 1960s, however, elements of inclusion of Hausa groups and other settlers into the larger urban society became important. First, the creation of local government in 1952 in the western region integrate into the municipal council a number of settlers that were initially discriminated by native authorities (Adeboye 2003, p. 310). Second, the growing domination of political parties in the 1950s and 1960s brought about a marginalisation of the traditional chiefs and facilitated the integration of Sabo residents into party machines. Unlike the Sabon Gari in Kano or the native town in Jos (see the following text), Nigeria's three main parties (NPC, AG and NPC) were operating in the Sabo district during the 1950s, an indication that there were internal divisions and that Hausa tradesmen were seeking forms of local or regional patronage (Cohen 1969, pp. 141–150). Third, the residents of Sabo converted *en masse* in the 1950s to the Niassene branch of the Tijjaniya order, which, contrary to Cohen's assertion, was not limited to Sabo residents, but was open to other groups, authorising the development of Islamic practices that transcended the ethnic affiliations of their followers during the 1950s and 1960s (Peel 2011). In other words, despite the colonial planning of divide and rule between natives and non-natives, political and social dynamics of the late colonial and early post-colonial periods helped integrating non-native groups into Ibadan society.

Exclusive forms of belonging to the city was not yet institutionalised nor reified. When the city became the seat of the south-western regional government in the 1950s, the need for qualified civil servants increased, but as the Ibadan elites preferred Koranic schooling to western education, they often lacked the training required to assume positions of responsibility. *Indigenes* of Ibadan (i.e. people claiming ancestral origins in Ibadan) were mainly employed at subaltern levels in the administration, for example, as gardeners, maintenance staff, or security personnel (Ajala 2008, pp. 160–162) while the position of governor was systematically occupied by *non-indigenes* until the Second Republic (1979–1983). Bola Ige, born in a small town at 120 km from Ibadan was the last governor elected against an *indigene* candidate in 1979. According to Ajala, it was during the Second Republic that Ibadan started to exclude *non-indigene* in Ibadan politics (Ajala 2008, p. 163).

The Federal Character promoted by the Constitution of the Second Republic has effectively triggered stronger 'claim over ownership': Ibadan *indigenes* increasingly claimed the right to hold political and administrative positions as well as ancestral origins in the city. Like in many other states in Nigeria, indigeneity has become key in gubernatorial elections especially with the return of democracy in 1999. Lamidi Adedibu, an indigene from Ibadan, became one of the major godfathers of the PDP presidential party against the party of Awolowo's successors by building political networks out of working-class associations representing urban populations with little education and who were excluded from the job market (Roelofs 2016, p. 102, see also Chapter 5). He campaigned in the 2003 election in favour of his godson a 'son of the soil' candidate (Rasheed Ladoja) coming from a neighbouring village of Ibadan against Lam Adesina, the incumbent governor, a *non-indigene* criticised by local associations for failing to protect *indigene* interests (Roelofs 2016, p. 97). Belonging to Ibadan has today become a political virtue: the two last governors (Abiola Ajimobi (2011–2019) and Seyi Makinde (elected in 2019) claim to be born in historical compounds of Oja Oba, the nineteenth-century city, a key element in the very definition of indigeneity in Ibadan. Indigeneity politics simultaneously became entrenched at the municipal level: between 1999 and 2005 only one *non-indigene* was elected as chairman among the eleven chairman of Ibadan local governments and among the 99 councillors from the local government councils only eleven were non-indigenes in 2005 (Ajala 2008, p. 163).

At the bureaucratic level, belonging to Ibadan has also become central. The indigeneity of the political personnel characterises the civil servants at every level of the state and local governments; their origins have been regularly examined since the return to civilian government in 1999. Oyo state has had several alternating governments since that date. Whenever there is a change of majority party, employees are asked to submit a series of documents including their certificate of *indigene*. Each change of party in power has the effect to eliminate from office people who were *non-indigenes*.[3] The exclusive presence of *indigenes* in local administrations is a naturalisation of the so-called ancestral rights. The strength of this discourse lies in being acknowledged today as a matter of course (Geschiere 2009, p. 5). In the interviews conducted at every level in the hierarchy of Akinyele and Ibadan North local governments, the exclusion of *non-indigenes* is seen as the norm; the directors of the departments interviewed (health, environment, public works, community services) in the two local governments admit that they do not employ *non-indigene* staff: it has been naturalised and is never presented as the result of a discriminatory policy.

This discrimination has become so fully incorporated into the local political and bureaucratic culture that it is not really challenged by the *non-indigenes*. Representatives of the Hausa community in Ibadan tried to request a more inclusive system from the governor of Oyo state:

'In 2001, we went to see the governor. I don't want to give his name. We went there with a man who was 102 years old and a woman who was 108, who were Hausa, both of them born in Ibadan. We told him: 'We have been here for decades, but we cannot secure jobs in the government'. Do you know what he told us? He said: 'There are too many people trying to secure jobs in the government and many people are settling in Ibadan. If we say that we are going to employ one single *non-indigene*, a hundred *indigenes* will come to complain'. We (as Hausa) have struggled to have democracy in Nigeria, but access to the civil service remains very difficult for us'.[4]

The governor in question, Lam Adesina (1999–2003) despite this discourse was perceived as not defending sufficiently the rights and interests of Ibadan *indigenes*. Gaining access to positions of responsibility as well as to subaltern positions has become almost impossible. In 2012, the Hausa representative in the Akinyele local government acknowledges: 'As it is very hard to enter public service [...], we do not even try to obtain certificates of *indigene*. We know that we will not be able to obtain them. We don't request them.

We don't even make the effort to seek one'.[5] Despite this, he still refutes the restrictive use of the term *indigene*:

> No matter where a Hausa community resides, the people see themselves as belonging to the host community. Among all the ethnicities of Nigeria, the Hausa are the only ones who construct a cemetery and are buried there. We do not take our dead to be buried in our villages of origin. It has been this way for centuries. So we are *indigenes* and this city is our city. Constitutionally we are not *non-indigene* because in every place in this country, Nigerians are free to do as they wish. But traditionally we are *non-indigenes*: we cannot apply for the position of *obaship*, we cannot become *bale*, etc. People call us *non-indigenes*, but we think that Ibadan belongs to us, we think we are *indigenes*.[6]

This leader knows that his definition of *indigene* (belonging to the place where one has settled) challenge the dominant view of political leaders, bureaucrats and more ordinary citizens claiming ancestral roots in Ibadan. Moreover, he and other Hausa leaders have been integrated into political parties and Sabo neighbourhoods have benefitted from the patronage of key political leaders like Adedibu. In other words, *indigenes* have succeeded, in the past 30 years, to monopolise administrative and political positions and to impose a narrative of urban belonging that is not fundamentally contested. Unlike in Jos.

Jos has a common colonial history of divide and rule with Ibadan but has witnessed a radical different trajectory from the 1990s. It is the capital city of the central state of Plateau, an ethnically diverse zone made mostly of minority ethnic groups running across central Nigeria. This mining town created by the British in 1915 attracted a huge flow of migrants both from the surrounding Plateau and from other parts of Nigeria. As the city continued to grow in the decades after Nigeria's independence, tensions and mutual suspicion have developed between locally based Plateau ethnic minority groups, such as the Berom, the Afizere and the Anaguta referred to in the literature as *indigenes* and groups portrayed as migrants (Hausa, Fulani or groups coming from Southern Nigeria) presented as 'settlers' or as *non-indigenes*.

The tensions between these groups culminated in violent conflicts at the turn of the millennium. The first erupted on September 2001 and started around the contested nomination of a civil servant from Hausa origin in a local government position of Jos North. The scale of massacres (death of more than 1000 people in a week, destruction of hundreds of houses, schools, shops, mosques and churches) is unique in the history of the city. It was followed by another one in November 2008 during local government elections in which at least 800 people were killed and again in January 2010 where violence broke out mostly in the rural areas on the edge of Jos. The scale and intensity of the successive riots is astonishing (around 4,000 and 5,000 people were killed between 2001 and 2010) and are seen by academics as one of the most atrocious in Nigeria's modern history (Madueke and Vermeulen 2018: 39). Specialists agreed that the conflict over indigeneity took a religious dimension in successive riots. There were mass killing of Christians in Muslim-dominated areas and mass killing of Muslims in Christian-dominated areas and 'mobs' moved around the city identifying, killing and maiming anyone of the opposite faith (Higazi 2007; Krause 2011). This violent antagonism was rather new in the 2000s.[7] According to Kingsley Madueke (2019), it is partly due to the controversial creation of one local government in 1991 which increased competition among groups and contestation over who is an *indigene* of the city. The threefold claim of belonging (belonging to a city, governing the city, claiming ancestral origin in the city) has become in the two past decades one of the most burning political issues in Jos.

Like elsewhere in Nigeria, colonial planners divided up the mining town into three distinct parts: the Jos township made for Europeans separated from Asian and other migrants from Southern Nigeria (Igbo and Yoruba), the Jos Native Town for migrants mainly from Hausa origins coming from Northern Nigeria and surrounding villages of natives some of which came to be included in the city with urban growth (Higazi 2007, pp. 75–76; Madueke 2017). If this planning pattern does not really differ from other colonial cities in Nigeria, competing historical narratives have been developing around the very foundation of the town. Hausa were maybe the first to settle in Jos: before the official foundation of the city, the place received the designation 'Hausa Settlement Jos' in 1912 (Plotnicov 1967, p. 41). But local populations – the Berom, Afizere and Anaguta – have been living in the area of Plateau for centuries. During and after the colonial period, Jos was attractive for migrants from the North especially for *talakawa*, the subaltern classes of Northern Emirates who wanted to escape domination of aristocratic classes; many of them have not kept relationships with the Emirates and developed instead a strong city belonging (Higazi 2007, p. 75). Populations from the neighbouring villages were not allowed to live in the Native town migrants and this had the effect of reifying ethnic boundaries and reinforcing mutual exclusivity (Madueke 2017). Moreover, the colonial administration gave the Hausa population a representative of the Native Town from 1914 to 1952 who took the title of Sarkin Jos, chief of Jos (Plotnicov 1967, p. 46). Unlike in Ibadan, the creation of a local government in 1952 did not ease relationships between groups. The colonial administration transferred power to the Berom and led to disputes over who was supposed to have authority over the city

(Plotnicov 1967, pp. 47–48). Antagonism was also fuelled by new political parties in Jos that were mainly based on ethnic affiliations.

This long colonial reified ethnic division was sharply reactivated in the 1990s. Military government created state and local governments to legitimise their action but the creation, in 1991, of one local government in Jos sharpened political competition. The unique local authority in Jos, the Jos Local Government Area (LGA) was divided into two local councils: Jos South and Jos North local governments. Jos North included the central city where the former Jos LGA headquarters, the main market, the University of Jos and the Plateau State's economic nerve centre were located. This creation guaranteed the Hausa population, which was the majority in the area but a minority in the city, the opportunity to lead the richest local government in the state (Madueke 2017; Ostien 2009; Higazi 2007). These new territorial and institutional boundaries were perceived by the three other minorities as a plan from the federal government to marginalise them. From then on, they adopted a stiffer stance fighting any government decision that appeared to favour the Hausa (Madueke 2017). The years following Jos North LGA's creation witnessed a proliferation of associations and other coalitions that actually worked to build group solidarity by emphasising ethnic and religious differences in the city through meetings, socio-cultural festivities and awareness-raising via literature distribution and rallies (Madueke 2018). Certifying who was an *indigene* of Jos became a burning issue as elected politicians were trying to impose their own understanding of urban ancestry to Jos. The first elected chairman of Jos North local government, Sama'ila Mohammed issued certificates of indigenes to Hausa population[8] but with the return of civilian rule in 1999, the new state governor, Joshua Dariye (a Christian indigene to the state), and the new chairman of Jos North's government, Frank Bagudu Tardy (an Anaguta Christian indigene), refuses to grant them certificates (Madueke 2018, p. 94). Certifying who is a true indigene of Jos has fuelled the successive episode of violence between 2001 and 2010 and has been seen by all political leaders in Jos North as a contentious problem to keep the fragile peace between groups since then.

While in Ibadan *indigenes* came to control the bureaucratic and political structures of the city, and everyone has agreed on who is entitled to have a certificate of *indigenes*, in Jos, claim over ownership has become one of the most important political disputes in the past twenty years. Groups violently disagreed on who is supposed to have the authority of the city, who belong to the city, and what does it mean to have ancestral origins in the city. This new contested urban belonging had a significant impact on the working of state and local governments. Jos North local government has become the most contested place in the city. Between 2002 and 2008, the chairman was assisted by a Hausa vice-president and, from 2015, the new governing coalition of the Plateau state imposed a power-sharing agreement in an attempt to pacify the situation: the position of chairman is held by one of the three minority groups; that of secretary is held by a Hausa.[9] These solutions are far from solving the problem and, on the contrary, reproduce suspicions within the local government and the trafficking of certificates within the government itself (see the following text). The organisation of chiefs working under the local administration have also been transformed. Only chiefs coming from the *indigene* groups have been recognised and paid by the local administration. All other chiefs are considered illegal.[10] The question of who is a true *indigene* of Jos has eventually become central in the dispute. The negation of the ancestral origin of Hausa population in Jos is repeated over and over again by ethnic minority groups in power in the local government: 'the Hausa populations came from the north of the country, they enslaved the populations of the Plateau during the period of the Sokoto caliphate, they only settled in the Plateau region with the arrival of the British'.[11] This narrative is used as a means of saving scarce public employment opportunities for those *indigenes* who claim that they do not have any other local governments to access the resources of the state.[12] Denying Hausa populations their indigeneity is understood by the latter as a political manipulation that wants to exclude them from these same resources. This new urban belonging obsession is mirrored in the transformations of the shape and layout of the certificates produced in the past two decades.

Producing Certificates, Identifying Urban Ancestry

The federal character of the regime and the significant role left to local governments in defining indigeneity lead to the existence of as many certificates as local governments. The procedure is radically different from one place to another: it can take five minutes or several weeks, it can require many steps or be a mere formality. These differences reflect, first and foremost, the varying importance given to indigeneity in the country's different administrations. The more the procedure is subjected to multiple verifications, the more obtaining a certificate reflects a political tension around the very definition of indigeneity. On the other side, the easier it is, the more it reflects the banalisation of a bureaucratic mechanism that is nonetheless discriminatory. The production of administrative documents must be situated in its social, political and institutional contexts and constraints while the graphic artefact of the documents indicates this new form of urban belonging.

In Ibadan, the applicant for a certificate does not need an identity document: he or she only needs to be able to answer a series of questions about the origins of his/her family. In each local administration of Oyo state, four or five lower echelon civil servants are responsible for conducting interviews with candidates for certificates under the authority of the civil servant in charge of the local government's administrative affairs. The interview takes place in Yoruba language and consists in verifying the authenticity of the narrative and confirming or invalidating the candidate's ancestry or origin. That is why the two terms – certificate of *indigene* and certificate of origin – are both used in administrative and everyday language. Candidates are asked an almost similar set of questions in Ibadan north local government:

'First of all, you state your family name. We know the common family names of the true Ibadan man. Next you must know the name of your family compound (*ile*) and its exact location (the street name is not enough). The true Ibadan man has always associated his family compound with a village on the outskirts of Ibadan. In the nineteenth century, those villages were the outposts of the warlords (*mogaji*) of Ibadan. The warlords or peasants who lived there were the dependents of the *mogaji*. Thus, all the people who originated in Ibadan have a village of origin, even those that were integrated into a city, such as Agodi, Agbowo, Apata, and Sasa. Finally, you must know the name of the Olubadan'.[13]

If the civil servant does not find the replies convincing, more difficult questions are asked to thwart those who want to pass for *indigenes*. The civil servants then ask:

'What are the main traditional feast days in Ibadan? Where is the sanctuary of Okebadan? Could you recite a few verses of the *oriki* of your lineage?'

If they still have doubts, a witness close to the applicant, most often the candidate's father, is called to the telephone. A written attestation from the head of the compound (the *mogaji*) may also be required to determine whether the applicant's family indeed originated in the compound concerned. This lengthy procedure implies that the applicant must have solid knowledge about his or her family and lineage. The civil servants assert they are upholding our mission:

'People are ignorant of their history. When they come here, we have to educate them; they have to be sent to their *mogaji* (head of the compound) who will tell them where they come from'.[14]

The applicant can obtain the certificate in a day if he/she is able to answer the very first questions of the officials: the family name of the father must be identified as a 'true Ibadan name', the applicant must know the name of the compound of the ancestors of her/his father (*ile*) as well as the name of the village attached to this compound's ancestor. In other words, the boundaries of indigeneity are defined by the family name and the dual urban and rural residence of the ancestor, a specificity of Ibadan. While dual residence seems to be entrenched in nineteenth-century urbanism of the city, the idea of associating a name and a place of residence is rather a contemporary use of historical ancestry.

The city was founded in 1829 as a military camp by several dozen warlords that were displaced from neighbouring cities and welcomed throughout the nineteenth century tens of thousands of military retinues, followers, family members and slaves, a very peculiar form of indigenous forms of urbanism and government. The city did not emerge out of a coordinated project managed by an overarching authority as had been portrayed in earlier academic research. Instead, the conflicting oral histories collected by Ruth Watson (2003) clearly demonstrate that warrior–founder compounds were constantly changing as new refugees arrived, built up their military retinues, and then departed to set up their own *ile*, taking their own group of followers and their families with them. Power was based on individual military merit and on the capacity of the warlords to accommodate people. When a nineteenth century warrior chief granted land to members of a departing party, there were still 'his people', he was simply expanding his political jurisdiction by settling them elsewhere (Watson 2003, p. 27). Many slaves and dependents from the warlord worked on farms at the periphery of Ibadan. This explains the dual residence required today to obtain the certificate of origin.

Identifying an Ibadan name and associating it to a specific compound (*ile*) seems more recent, however. Watson rightly insists against a trend of anthropological and historical literature on Ibadan: the social/spatial boundaries of lineages, compounds and quarters were so volatile that it is impossible today to identify a primary group (Watson 2003, p. 28; see also Adeboye 2003, p. 305). The place of compound in identifying an indigene was probably less important before. In the early colonial period, the definition of an indigene was a person born in Ibadan to an Ibadan man or related by blood to an Ibadan man, no specific place was then mentioned (Falola 1985, pp. 54–55). Moreover, in the 1970s and in the 1990s there were a number of court cases on competing claims to the post of *mogaji*, a term that came to designate the head of Ibadan *ile* from the 1860 onward (Watson 2003, pp. 25–30). These court cases indicate that the process of associating a compound and a name was very much contested. Identifying an *indigene* of Ibadan is not as easy as bureaucrats claimed. To refer to the current heads of compound to check whether the applicant is a descendant of his compound gave him the power to legitimise his own ancestral claims and to exclude other claims.

The politicisation of ancestral claims is more a legal issue than a bureaucratic one, however. For the civil servants, these narratives are instrumental in developing a civic sense and asserting local citizenship. Many candidates are surprised, disappointed and discouraged to discover that the certificate cannot be obtained without being able to answer this series of questions. They are allowed to return home, however, to ask their parents and elders of their lineage for precise information before submitting another application. Bureaucracts see this process as a way of developing the civic awareness of citizens and encouraging them to learn their local history. While candidates must know their history to claim their origin, civil servants must have precise, diversified and relatively exhaustive knowledge of the history and politics of their district. The Ibadan North local government area has hundreds of compounds and the Akinyele local government has hundreds of villages, whereas the civil servants merely have a list of villages without any other available documentation.[15] They were trained by watching and listening to their supervisor conduct interviews. The daily conversations between candidates and these civil servants have enabled them to acquire an impressive body of local historical knowledge. In the process, they become historians of a particular kind, relying solely on their memory to distinguish true *indigenes* from the false. Although common knowledge of local history is the core of the identification procedure, these civil servants occupy a position of authority: the candidate does not know how much the bureaucrat knows from the set of ready-made questions. If they are not sure of the answer – for example, if the name of the applicant or of the compound means nothing to them – they can demand additional informations. The whole process is obviously perceived differently by civil servants and users. Local civil servants see it as strengthening the feeling of belonging to a political community, because it implies that candidates know the history of their family and their community. Candidates, on the other hand, usually experience the process as arbitrary, time-consuming and predatory, even though certificates are acquired through a procedure based on standardised questions.

Despite this, no one in the office contest the very ways to identify *indigene* nor the rights that go with them in Ibadan. The procedure is not always that rigid. The process is first open to women who are not from the area, who may become *indigenes* and request a certificate if they take their spouse's name. This is in contradiction with the recommendations of the Federal Character Commission prohibiting women to change their status once married to avoid the claiming of dual indigeneship (Manby and Momoh, 2020, p. 14–15).[16] It could also be opened in a very discretionary manner to *non-indigene* groups of Yoruba origin. In Akinyele local government office, not far from the federal university of Ibadan, civil servants welcome several thousands of students during university admission, between August and October. The long queues of students in front of the local government offices *de facto* promote the 'sale' by local intermediaries of narratives that are in line with the expectations of the officials in charge of verifying the authenticity of the demands, thus allowing revenues into the budget of the local government (Fourchard 2015, 50–52). In the vast majority of cases, applicants for a certificate have to comply with local demands. A laxer verification of narratives in times of high demand does not prevent the same officials from thoroughly reviewing applications in off-peak periods. In Akinyele, a student was unable to name the village of origin of her father's lineage and the characteristics of that village. She had to leave without the certificate, the official telling her to enquire with her father.[17] To a father who came to ask for a certificate for his daughter living in Lagos, he replied that she had to collect the certificate in person. In other words, if the discretionary power participates in the issuance of these documents, and if the rules can at times be left to the discretion of officials at the office, these norms are nonetheless highly bureaucratised and routinised. Negotiating a certificate is possible under certain conditions and for some individuals; however, for the majority of citizens that are not known by the administration, strict rules are applied in Ibadan but also in Jos and Lagos (Fourchard 2018).

These rules show the importance of the state production of documents in the reproduction and creation of social differences among the population of citizens (Sharma and Gupta 2006: 17). The procedure to get a certificate is not contested; there is instead an individual appropriation of bureaucratic procedures a process referred as *eigesinn* by Lüdkte. In 2011, a scheme known as the YES-O programme (Youth Employment Scheme of Oyo State) offered 20,000 jobs to unemployed youth in the state. In a few weeks, 180 000 youth had applied for a job. The government did not mention that the plan was designed for *indigenes* but local governments were faced with a rush of applicants seeking certificates (Fourchard 2015, p. 49). Indigeneity is so ingrained in the public consciousness that applicants thought the scheme was for indigenes only. This reveals how dominant indigeneity narrative and dominant bureaucratic procedures have been appropriated by citizens in the past three decades. These concrete situations involving the practical exercise of power – the *douceurs insidieuses* of Michel Foucault – translates into the desire to receive social or economic benefits (Hibou 2011, pp. 26–33; Lüdkte 2015a). In Ibadan political and bureaucratic dominant norms on indigeneity are not contested but might be ignored (see the following text), unlike in Jos where getting a certificate has largely fuelled the conflict between groups.

Hausa population in Jos were provided with certificates of *indigene* in the 1990s but have been excluded from this possibility since the election of a new Christian *indigene* governor and a new chairman in Jos North local government in 1999. Since then, to get a certificate is the object of a tough bureaucratic procedure controlled by numerous key actors, which has led to an ethnicisation of the document: certificates are now reserved to only one of the three ethnic groups who claimed to be the only *indigenes* of Jos. To get a certificate of indigene is also perceived as a political act for the Hausa population of Jos who are excluded from this procedure but who considered themselves as full citizens and founders of the city as they named themselves Jasawa or people of Jos. In between the antagonistic groups, the production of certificates has become the object of considerable political and bureaucratic anxiety especially in Jos North local government.

To obtain a certificate in Jos, the procedure is much more restrictive. It requires meeting with six intermediaries. The applicant starts by passing through the offices of the local government to pick up an indigene identification document given by an official who, since 2001, only grants it to applicants belonging to one of the three ethnic minority groups (Afizere, Berom or Anaguta). This is the first classification to start with. The form must then be signed by the ward chief, the village chief and the district chief fully recognised by the local administration. This is the second verification. The applicant finally returns to the same local government official who checks the quality of the signatures and stamps, issues the indigene certificate, which has to be signed and stamped by the president of the local government and again by the district chief. This is the third and last verification. The process can take many weeks and depend on the availability of chiefs and officials. As mentioned by Matthew Hull (2012a, p. 12), graphic artefacts are not neutral purveyors of discourse or simply the instruments of already existing social organisations. Their specific material forms precipitate the formation of shifting networks and groups of official and non-official people and things. In Jos, the long chain of verifications is an old practice that predates the 1990s tensions,[18] but the micropolitics of the artefact has changed since the 2001 violence.

The conflict imposed small yet extremely significant changes to the certificate. There are three of them mainly meant for excluding Hausa to get one (Figure 6.1). First, before 2001, a simple mention of ethnic belonging was required. Since 2001, it is necessary to belong to one of the three mentioned groups (Afizere, Berom or Anaguta). This alteration is a direct consequence of the crisis on the content of the document. In the neighbouring rural local government (Bassa) that was spared of the violence until 2016, the certificates

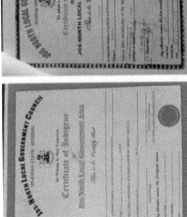
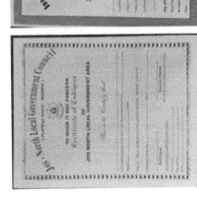

FIGURE 6.1 Certificates of Indigene in Jos North Local Government (in 2015, 2016 and 2017).

have not been modified and are similar to those that were produced in Jos North before 2001: ethnicity is required but does not condition access to the certificate. This rural local government was until 2018 untouched by this process of ethnicisation that developed in Jos North: three Hausa clans representing several hundred individuals living in the district since the nineteenth century are actually recognised as *indigenes*. The very violent crisis in Jos has precipitated the exclusion of any ethnic groups that do not belong to the three identified groups. Second, in most neighbouring local governments, the secretary is authorised to sign the certificate, especially if the president is absent. But in Jos North, since 1999, only the chairman has the authority to do so, *de facto* excluding other administrative officials (the vice-president or secretary). This is so because the vice-president and the secretary, who were Hausa between 2002 and 2008 and since 2015 are suspected of distributing certificates of indigenes to their fellows. The third modification is related to the chieftaincy. A countersignature of the district chief of the local government on the certificate is required. However, this position of district chief does not exist today in the Hausa hierarchy of Jos.

The crisis situation fuels a climate of suspicion within the local government since the power-sharing agreement imposed by the state governor in 2015 (the chairman being from one minority groups, the secretary being from Hausa origin). In November 2016, the secretary was suspected by the president of the local government of trafficking certificates in favour of members of his group. According to her, the secretary had documents that he 'sold under the cloak' and that he signed with the name of the local government. This led the chairman to change the certificate format between 2015 and 2017 (Figure 6.1). Behind the aesthetics of the new document (red stripe, writing in different colours, red frame where the stamp of the president of the local government must go), the idea is to make it more difficult to reproduce. In January 2018, the new chairman informed me of the existence of trafficking of false documents run by the new secretary of the local government. His administration has once again undertaken the production of a certificate with characteristics (he insists on the paperweight) that would prevent its falsification. The alteration of the certificate is always accompanied by a change of identification form. These efforts reflect the 'bureaucratic anxiety' of local government officials faced with a political conflict they find themselves involved in (Hull 2012b, p. 255) as they are trying to deal with by strengthening bureaucratic controls.

Documents are effectively produced within the local governments by the secretary benefitting the Hausa populations. This could easily be qualified as illegal or forged. However, they are above all a pragmatic response to the techniques of exclusion deployed by the local government. On a paper with letterhead, logo and address of the local government, and just like very common administrative documents, is indicated 'to whom it may concern' followed by the family name, then, it is mentioned 'the name above is an indigene of Jos North'. The document is signed by the secretary of the local government and the relevant ward chief for the applicant.[19] The document looks exactly like a document attesting the indigenous origin of its holder, but it is actually very different from the certificate produced by the local government. In fact, it looks more like an administrative letter. Nevertheless, the document might be perceived as reliable by administrations in other parts of the country. It can easily be presented to a university or to the recruitment service of the army that would not easily recognise a fake one, given the multiplicity of types of certificates circulating throughout the country.

The secretary of the local government makes no secret of the fact that he produces certificates on behalf of the local government. He even confirms that this is an old practice linked to the presence of a Hausa leader in the local government as vice-president (2002–2008) or as secretary (2015–2018).[20] For him, it is above all a political issue: 'Hausa populations are indigenes of Jos, they are in the majority in the local government, so they have no other reason than a political one to be excluded from the right to obtain a certificate'.[21] In this context, producing a certificate corresponds to a political act, one of bureaucratic dissent. For his part, the chairman of the local government knows that the issue of access to the certificate is at the heart of the conflict in Jos; he knows that what his secretary does is probably illegal, yet he does not report it to the police.[22] In such a volatile situation, reporting the existence of a parallel trafficking within the local government and trying to put an end to it by dismissing the secretary or bringing him to court is likely to trigger further violence. Rather than take such a hazardous decision, the chairman chooses a cautious bureaucratic solution: to produce new types of certificates every year in the hope that this inflation in documentary creativity may dry up the trafficking of fake certificates.

Indigeneity, Segregation and Patronage

The level of mass violence that has been occurring in Jos for more than a decade has changed the socio-spatial pattern of Jos. Neighbourhood-based segregation has been reinforced along ethnic (*indigenes* versus *non-indigenes)* and religious lines (Christians versus Muslims). In Ibadan, the two neighbourhoods inhabited by the Hausa population are not areas discriminated in terms of access to service for different historical reasons that need to be explored. Quality of services provided by the local governments and different patronage networks play a more determinant role in side-lining the exclusion of Hausa population from public offices and mitigate discrimination.

The successive episodes of mass violence in Jos affected neighbourhoods differently according to religious belonging (violent events occur in ethnically mixed areas while armed mobs often originate in segregated neighbourhoods) and according to income (middle-class areas received more police protection and were easier to police than overcrowded slum areas (Krause 2011; Madueke 2018). In some cases, action by religious leaders were key to limit violence (Krause 2017). Almost everywhere, however, the violence reinforced segregation between Christian and Muslim neighbourhoods while mix neighbourhoods were reduced in number. After the 2001 riots, the central neighbourhood of Angwa Rogo-Angwan Rimi has lost its minority of Christian residents (34 per cent in 1995), the population became homogeneously Muslim with over 100 mosques and Koranic schools; the only two churches that once stood were both destroyed in the violence (Madueke 2018). Muslims have equally been forced to vacate neighbourhoods like Eto Baba and Jenta (Krause 2011, p. 36). Ethnic composition of residents' associations, youth and vigilante groups, and other neighbourhood-based organisations have also become homogeneous entirely comprising co-ethnics and co-religious (Madueke 2019). This contributes to exacerbate collective violence in the following riots. In 2008, both segregated and mixed settlements contributed to violence but armed mobs were likelier to originate from segregated neighbourhoods that could mobilise large-scale and ready-to-fight groups while mixed settlements, especially those sandwiched between rivalling segregated ones served as frontiers for fighting (Madueke 2019).

Before 2001, vigilante groups were combatting, property-related delinquency and interpersonal violence common in Jos, in Ibadan or Lagos (Lar 2015, Chapters 3 and 4). The successive riots came to redefine their role and their presence in many neighbourhoods. They were by far the only actors in the successive Jos riots but in many cases, youth associations and vigilantes were made responsible for patrolling and preventing troublemakers from gaining entry. Vigilante groups turned back to duties of controlling crime and patrol after the riots but were actually transformed by the very experience of violence (Lar 2015). As boundaries between Christian and Muslims neighbourhoods hardened in a post-violence context, vigilante groups became also more homogeneous in neighbourhoods affected by violence. The district head of the Christian area of Kabong interviewed by Jimam Lar narrates how vigilante policing seized in his area following the outbreak of violence in 2008. When he was invited by the police to set up a new vigilante group, he succeeded in recruiting 47 boys across the district, who were strictly 'Christians boys', as Muslim boys have all left the area because of the violence.[23]

Once vigilante groups turned into a militia, breakdown of exchange relations make it hard to re-establish the trust between divided neighbourhoods (Higazi 2016, p. 373) while homogeneous ethno-religious patrols contribute to feeding mutual suspicion between neighbourhoods.

The past 20 years of conflicts of indigeneity have been shaped by competing claims over ownership in Jos. First, one of the direct consequences of the 1991 gerrymandering process of allocating a specific urban space to a specific group (Hausa) led to a non-ending contestation over who has the authority to govern Jos North local government: the nomination of a Hausa local government official in Jos North was the direct cause of the September 2001 riots; the election of a new local government dominated by Hausa was cancelled in 2008 and triggered a new wave of violence; and the planning of the 2019 local government election was stopped because rival groups were ready to fight if the Hausa majority of voters were going to win the election. Second, this politics of indigeneity and successive episodes of violence has reinforced the increasing spatial division of Jos along religious and ethnic lines. Belonging to the city increasingly means today belonging to his or her own neighbourhoods and to associations and groups composed of the same co-ethnic and co-religious fellows. And finally, the ancestral origin in the city is an unfinished dispute between rival groups claiming to have arrived first in Jos. This conflict is far from over and is instead manifested by the attempts at complicating the access to certificates of indigene and by the parallel production of alternative certificates within the local government office itself. This situation without being exceptional represents fortunately a minority of cases in Nigeria. In many cities like in Ibadan, *indigene–non-indigene* relationships are shaped by negotiated relationships despite open discrimination.

Excluding *non-indigene* from access to political and bureaucratic had not radically changed the shape of the city of Ibadan. To explore the possible spatial effects of the politics of exclusion of non-indigenes in Ibadan, I have looked at two neighbourhoods inhabited mainly by populations of Hausa origin : Sabo, a colonial township exclusively made for the Hausa population and Sasa, a neighbourhood that accommodated Yoruba farmers in the late 1960s, and Hausa traders forced to move their residence from the centre of Ibadan during the 1970s. Although the populations of these two neighbourhoods are similar, their level of development is not the same. Sabo has a decent level of infrastructures and services, above the average in the metropolis, whereas Sasa, originally a well-planned housing estate, has turned into a slum over the last thirty years. In 2003, I attributed the particular trajectory of this neighbourhood to discrimination against the Hausa population with regard to property and to the local government's deliberate neglect of an environment presumed to be in the hands of the Hausa merchant community

(Fourchard 2003). But new sources have allowed me to revise my initial interpretation and show how the different forms of patronage cultivated in the two neighbourhoods offer a more relevant key to analysing them, which downplays the discriminatory aspect of public action in accounting for their development.

In Ibadan, discrimination against *non-indigenes* and the preference given to *indigenes* has not led to collective protest. In Sabo, these devices are either ignored, perceived as marginal or circumvented by clientelist relationships. For many traders, having a certificate is a marginal matter. Many of them have never even heard of it, particularly those who have not been to university. Furthermore, there seems to be a relative acceptance of this dual citizenship. As one local political leader said, 'After all, if I want a certificate, all I have to do is go back to my local government to get one; if I want to go to university, I can choose between enrolling in a state university in the North where I come from, or enrolling in a university of Oyo State if I can afford it'.[24] As for the new employment programme in the state (the YES-O programme), well-established traders suggest that the young people in the neighbourhood were not interested, not because they were wilfully excluded, but because the jobs on offer were poorly paid and unappealing. In short, the perception of discrimination in gaining access to employment and the university is mitigated by the idea that discrimination against *non-indigenes* is standard practice on a countrywide scale.

Above all, the perception of discrimination is attenuated by the absence of discrimination in gaining access to public goods and services. While civil servants admit that public jobs are reserved for *indigenes*, they also point out that there is no discrimination in the services provided in *non-indigene* neighbourhoods. There is reason to believe them. On the whole, access to schools, health care, drinking water and electricity is better in the Sabo neighbourhood (where a very large majority of the population are Hausa) than in the historic centre of Ibadan (where most of the population say they are *indigenes*). The former is a well-equipped neighbourhood, whereas the latter is considered by the local government chairmen as a large slum due to the lack of basic services (access to water, electricity, schools and sanitation). In reality, there is a different rationale for urban planning – or lack of it. The transformation of the entire city-centre into a very poor area is the result of a combination of factors distinct from indigeneity politics: absence of action and planning of the old city to avoid alienating the 'traditional' authorities of the city; overpopulation linked to on-site housing densification; pauperisation of households since the cacao-based economy disappeared in the 1970s and the movement of the salaried/wage-earning population to better equipped neighbourhoods (Fourchard 2003).

On the other hand, the historic neighbourhood of Sabo, have benefitted the most from infrastructures in terms of roads, access to water and electric power since the colonial period. Today, numerous civil servants seem to suggest that the local governments have made significant efforts to improve *non-indigene* neighbourhoods, sometimes even more than in the other parts of the city. Civil servants in charge of the environment claim that large markets located in these neighbourhoods provide more work for sanitation teams, whereas public health officials assert that resistance to vaccination campaigns against polio is much greater in these areas and consequently demand that more energy and personnel be devoted to convincing the population that their intervention is justified, a process found in other cities of Nigeria (Yahya 2007). In fact, the level of infrastructures and services that are provided depends mainly on a set of networks of client–patron relationships between the governor, the chairmen of the local governments and their political clients. Resources are limited and civil servants readily admit that local politicians choose the electoral wards in which they want to invest. The heads of the various services submit a list of the work to be carried out to the president of the local government (digging wells, putting up electric pylons, asphalting roads, etc.). After consulting with the neighbourhood associations, the chairman chooses among all these projects, not on the basis of balancing the investments to be made in the metropolitan area but on the basis of the electoral support received in the different parts of the city. He then entrusts the decision as to which part of the neighbour-hood should be of benefit to the local leaders of his party.[25] The provision of services and the amount of public investment therefore depends on this form of patronage in which citizens' votes and clientelistic relationships are far more decisive than the division between *indigenes* and *non-indigenes*. In this regard, the highly unequal quality of infrastructures and services in the two neighbourhoods of Sasa and Sabo can be explained above all by the difference in the type of patronage operating within them. The woeful condition of the neighbourhood and market in Sasa is due less to a discriminatory policy against the Hausa than to the fact that patronage relations are much less favourable to the majority of the population.

During the 1990s, the Sarkin Sasa (the traditional authority that was supposed to represent the interests of the Hausa community in Sasa) became one of the most important clients and a confident of President Sani Abacha (1993–1998). They were both from Kano. They began spending time together when Abacha became commander-in-chief of the Second Armoured Division in Ibadan between 1983 and 1985, at a military camp near Sasa. He also became a powerful patron in Ibadan, redistributing money and consumer goods, particularly cars, to several political leaders of the metropolis and the state. Under the Fourth Republic, he was also involved in financing the

PDP, and, together with Lamidi Adedibu, helped to elect two governors from that party, one in 2003 and the other in 2007. However, unlike the main political godfather in Ibadan, who had solidly embedded supporters in his neighbourhood in Molete (Chapter 5), the Sarkin Sasa did not see the point in cultivating neighbourhood networks and, on the contrary, neglected the development of his own neighbourhood. He took the liberty of levying illegal taxes on market traders and paid no heed to the grievances of the association that opposed him.[26] He had a palace built in the neighbourhood, and despite the intense hostility of the traders, replaced the reservoir needed for the market with a police station, which he financed himself to benefit from permanent personal police protection. The local government had no other choice but to turn the levying of market taxes over to his henchmen, thereby depriving itself of revenue while the market seriously deteriorated.[27] In short, the decayed state of the market and the lack of investment in the neighbourhood were due to the ongoing battle between this patron and the traders, both Yoruba and Hausa, more than to a deliberate policy of discrimination carried out by the government against the *non-indigenes* of the neighbourhood as I had originally believed (Fourchard 2003).

In contrast, the residents of the Sabo neighbourhood benefitted from a generally more favourable form of patronage through the support of Adedibu, who reached an agreement with the National Electric Power Authority (NEPA) guaranteeing the neighbourhood a regular supply of electric power by paying the bill on time (Obadare 2007, p. 125). This was an important privilege on the scale of a metropolis frequently plunged into darkness and a rare service on the countrywide scale, as attested by the popular expressions used to describe the company's performance ('No Electrical Power Anytime' or 'Never Expect Power Again'). In this context, the neighbourhood largely voted for the PDP in the 2003 and 2007 elections. After Adedibu's death in 2008, the neighbourhood lost both its primary connection to the PDP patronage network and its electric power connection, which generated public discontent with the government. A few months prior to the election of the governor in 2011, an ACN MP in the opposition at the time decided to install a new power transformer in the neighbourhood to enable the return of electric lights, and according to local political leaders, to shift voters away from the PDP and over to the ACN.[28] In other words, political patronage is viewed by the residents of Sabo as a means to obtain more advantages than other neighbourhoods. These two examples show that possible discriminatory policies against *non-indigene* populations are in fact far less decisive than patronage and electoral clientelism, which, depending on the circumstances, can be used to give preference to some neighbourhoods and penalise others. In that regard, patronage today, like during the 1950s and 1960s help to bridge institutionalised attempt at reifying *indigene–non indigene* boundaries.

Conclusion

Studying certificates from a detailed description of the contexts in which they are issued shows that the boundaries of indigeneity are unevenly porous from one locality to another. The two empirical examples testify to the radical differences in what this term means and represents within different administrations and the ease or the difficulty with which a certificate can be obtained across the country. The line can be crossed under certain conditions in Oyo state, it is almost impossible to cross it in Plateau state due to persistent conflict and the stricter control of bureaucratic procedures and political control while it can be easily crossed in most local governments of Lagos (Fourchard 2018). This unequal access to local citizenship shows the importance of not considering indigeneity as a political problem in itself in Nigeria. Although the certificates signify an essential division between citizens, their production does not in itself lead to conflicts. Unlike many other countries that have experienced similar trajectories of identity closure, it is not the existence of an apparatus that excludes part of the population from access to state resources that is problematic, but rather its practical implementation that redefines the boundaries of local citizenship. Unlike what a large set of indigene-settler literature says, the comparative analysis clearly indicates that the production of the same documents is not necessarily a source of tension: if the certificate is at the heart of all the political battles in Jos, it is a marginal and even not a politicised matter in Ibadan, including for *non-indigene* citizens.

The production of certificates in local government offices actually reveal the power of bureaucratic domination over individuals. Discrimination against *non-indigenes* has become in three decades so entrenched in the history of the local administration that it has been naturalised. Issuing a certificate of origin is perceived by a number of civil servants and users as similar to issuing identity cards, birth certificates or driving licences, but a certificate of origin is not a document like any other, however. The procedures determine more or less sharply the borderline between *indigene* and *non-indigene* citizens and attest to the state of power relations between influential actors in a given society (between civil servants and politicians, between bureaucrats and chiefs and between ethnic communities). It gives rights that are denied to others and helps to fuel conflicts in numerous Nigerian states or local governments. Although indigeneity is an explosive issue in a number of places like Jos, the populations of many other cities or states in the country are indifferent to these forms of discrimination, adapt to them or even accept them. The 'insidious gentleness' of the bureaucratisation of indigeneity – the production of certificates giving rights to certain parts of the population at the expense of the others – have played a significant role in this process of normalisation, while constructing institutional boundaries between *indigenes* and *non-indigenes*. This *eigesinn* or the consent to such institutionalised discrimination is revealed by everyday practices: *indigenes* try to obtain a certificate to find employment or to pursue a career in academia, administration or politics. In Ibadan, the mass production and commodification of certificates, particularly in urban areas close to the offices that handle the most applicants, have the paradoxical effect of including more *non-indigenes* than civil servants are willing to admit but it does not challenge the overall system of discrimination against *non-indigenes*. In Ibadan, the dominant ancestral narratives only partly located in nineteenth-century urbanism are poorly contested: *non-indigenes* have no choice but to ignore it or circumvent it. In Jos, while the controversy is about who is entitled to have these certificates, nobody is asking for their even temporary suppression even if it fuels regular tensions among groups. Instead, different local government officials produce new forms of certificates and alternative certificates that testify the forces of bureaucratic graphic artefacts in this conflict-ridden society.

The urban and spatial effects of *indigene–non-indigene* relationships are fundamentally shaped by the hardening of the frontiers between the groups. In Jos, the cumulative effects of the politics of indigeneity, of issuing certificates and violence have strongly sharpened religious and ethnic divisions. In Ibadan conversely the perception of origin-based discrimination is tenuous or even non-existent, which partly invalidates the Human Rights Watch report on discrimination against *non-indigene* populations with regard to exclusion from basic services. The quality of those services is conditioned more by relationships and forms of patronage linking neighbourhood populations to state and federal government politicians. When state resources are used to consolidate the authority of a political patron, conflicts that could be interpreted as ethnic or as examples of the *indigene/non-indigene* divide actually have more to do with the abuse of power by those in positions of authority and protest against the unequal redistribution of resources.

Notes

1. The Ibadan North local government had 302,000 habitants according to the 1991 census and 308,000 according to the 2006 census; the Akinyele local government had 140,000 inhabitants in 1991 and 211,000 in 2006.

2. Section 42 of the constitution stipulates: 'A citizen of Nigeria of a particular community, ethnic group, place of origin, sex, religion, or political opinion, shall not (a) be subjected to disabilities or restrictions

to which citizens of Nigeria of other communities, ethnic groups, places of origin, sex, religion, or political opinion are not made subject; or (b) be accorded any privilege or advantage that is not accorded to citizens of Nigeria of other communities, ethnic groups, places of origin, sex, religion, or political opinion.

3. Akinyele local government interview with the administrative officer, 8 May 2012.

4. Interview with the representative of the Hausa community in Sasa, 4 May 2012.

5. Interview with the representative of the Hausa community in Sasa. 4 May 2012.

6. A representative of the Hausa community in the Akinyele local government, 4 May 2012.

7. Although there is a recent reinterpretation on how small-scale violence in 1994 was important in the radicalisation of the 2001 riots (Madueke 2019). NPC Hausa militants assisted by Berom participated to Igbo massacres in September and October 1966 in Jos prior to the civil war (Higazi 2007).

8. Interview secretary of Jos North Local government, 17 January 2018.

9. Interview chairman of local government Jos North, January 2018.

10. Interview secretary of local government Jos North, January 2018.

11. Stories reported by many interlocutors, including one district chief of Anaguta, Jos, October 2016, and the official in charge of issuing certificates in the local government of Jos North, January 2018.

12. Interview with Yakubu Bala Sisok, district chief and former elected official in the local government of Jos North, 28 November 2016.

13. In Akinyele, the questions are similar: surname, given name and location of the family compound, name of the *baale* and of the president of the local government, and the main characteristics of a place (church, school, mosque).

14. Interview with the civil servants of the Ibadan North local government.

15. The civil servants do not know the exact number of compounds in their local government. The Ibadan North local government has 12 neighbourhoods, some neighbourhoods have 30 compounds and a single compound can accommodate 500 people.

16. This practice does not seem to be systematic on a national scale.

17. Observation in the local government of Akinyele, Ibadan, May 2012.

18. Countersignature by the ward, village and district chiefs was already required in 1989 to obtain a certificate.

19. For obvious security reasons it is not possible to show this document.

20. Interview with the secretary of the local government in Jos North, January 17, 2018.

21. *Ibid.*

22. Interview with the president of the local government in Jos North, January 17, 2018.

23. Sallah Mandyeng Sha, interviewed by Jimam Lar, Kabong, Jos, October 2012. I would like to thank Jimam Lar for allowing me to reproduce this quotation.

24. Interview with a political leader in Sabo, 8 May 2012.

25. Interviews with civil servants, Ibadan North local government, 7 May 2012.

26. Interview with the traders' association in the Sasa neighbourhood, 10 May 2012.

27. Interview with the general secretary of the Akinyele local government and the local tax collectors at the market in Sasa, May 2012.

28. Interview with two ACN leaders from Sabo, 12 May 2012.

Conclusion
The Urban Legacy of Exclusion, Policing and Violence

Research that attempts to provide a detailed description of the actors' everyday life and show the connections between different historical periods and the various scales without sacrificing comparative analysis presents methodological, epistemological and political challenges. Yet *a posteriori*, this approach seems indispensable: it enables us to grasp how apparatuses and instruments of power, which were extremely uncertain, experimental and sometimes arose at random in the beginning, have reified, solidified and naturalised differences over time. This Foucauldian inspired genealogical analysis has revealed the classifying obsession of urban colonialism and its contemporary partly unexplored legacies: urban youths vs. adults, natives vs. non-natives, *indigenes* vs. *non-indigenes*, urbans vs. migrants from reserves, South Africans vs. international migrants. It demonstrates that these inherited forms of exclusion – which often intersect with racial differences social and gender inequalities – can also be violent, determining factors in the way social relations have developed and an expression of powerful relationships of domination. In retracing this genealogy, this book is first an attempt to reveal the distinctiveness of African urban histories in shaping global urbanism.

Contemporary post-colonial cities have a genealogy of colonial manufactured divisions that have yet to be explored in a more systematic way. If colonialism divided people to rule them, it remains unclear on what line this division became concretely operational in Africa's cities. The role of colonialism in sharpening the division between urban youth and adults was one of the central elements for governing an orderly colonial urban space. The category of the juvenile delinquent was invented in Lagos by the first social welfare service of the British Empire when the adaptation of certain youth and children to the urban environment was found to be problematic. The wish to produce a stable, autonomous and responsible male working class became simultaneously a powerful standard against which an increasing

Classify, Exclude, Police: Urban Lives in South Africa and Nigeria, English Language First Edition. Laurent Fourchard.
© 2021 John Wiley & Sons Ltd. Published 2021 by John Wiley & Sons Ltd.

part of the population of 'unemployed', 'vagrant', 'unruly' urban youth and 'single' women were criminalised. The extension of a state at once compassionate, coercive and intrusive into the lives of urban families resulted in collective penalties for minors (corporal punishment, deportation to the countryside and fines) based on exceptional penal regimes that were no longer applied to adults in the 1960s–1970s. This was largely supported by a large section of African urban societies from traditional chiefs, politicians, civil servants and vigilante groups that participate in the deportation of wandering youth, isolated women or undesirable poor to the countryside (Burton 2005; Glaser 2000; Waller 2006). Several independent states committed to nation-building glorified wage labour and kept this past coercive law and practices in their legislation (Barchiesi 2019, p. 64). Laws against 'vagrancy' were passed to maintain involuntary labour to target young rural population fleeing 'the bush' to seeking refuge in cities in Congo Brazzaville, Senegal and Madagascar (Keese 2014, p. 393). Deporting large sections of undesirable people to the countryside is part of a common practice as shown by the labelling names or operation set by national and local authorities: the *encombrements humains* targeting the vagrants, the *talibe*, the prostitutes in Dakar, the *wuhani* or the young and unproductive poor in Dar es Salaam, 'professional or able-beggars', i.e. unemployed youth and occasional street vendors in Kano because they were a stain on the reputation of the northern region, prostitutes, marginals and 'unproductive people' in Maputo after independence of Mozambique or more recently the informal workers in Harare under the operation Murambatsvina or 'restore order' (Burton 2007; Faye and Thioub 2003; Fourchard 2018; Morton 2019; Potts 2006). In South Africa, government intrusion took place on such a massive scale that several generations of children and urban youth underwent violent socialisation at the hands of the very institutions set up to protect them. This role of massive incarceration in South Africa is for a number of researchers central in the reproduction of violence in townships and in South Africa at large.

If colonialism ruled by division, it is unclear what the socio-spatial effects were in urban areas of such divided *apparatus*, beyond the racial division. There is a long, unfinished and controversial debate in African studies on the pervasive effects of colonialism in the manufacture of ethnicity. Some suggest that ethnic belonging was largely a colonial creation due to the multiple interventions of colonial administrators, European ethnographers, missionaries, traditional chiefs or catechists (Amselle and M'bokolo 1985; Chrétien and Prunier 2003) while many others suggest that ethnicity was not 'invented' but part of a longer precolonial history that colonialism only modified and reified (Lentz and Nugent 2000; Reid 2011; Spear 2003). This reflection is careful in examining the moving frontier of ethnic belonging, and if the urban fabric was sometimes acknowledged in pioneer historical and anthropological studies, it seems to be, however, marginal in today's debate. Some sections of urban studies especially from the 1990s have instead focused on the central place of European racism, of the civilising mission and colonial surveillance in the making of a colonial dual city and the resistance of the colonised people to this totalising ambition. Both academic traditions insisted on the classifying obsession of administrations but the role of colonialism in shaping a more complex socio-spatial urban fabric of Africa's cities remains unclear.

Racial delusion of colonial power was in urban areas articulated to classification based on age and sex as mentioned before but simultaneously on the origin of population living in cities. The incessant mobility of the population between the city and the country has remained a ubiquitous obsession of colonial administration in Africa not because the apartheid or the colonial city 'was not made for Africans' as it is sometimes suggested in geographical literature but instead because urban spaces accommodated different segments of workers according to different needs of the colonial economy. The creation of new urban enclaves in Nigeria in the early twentieth century exclusively reserved to 'migrants' or 'non-natives' and separated from other residents and the systematic and pervasive distinction between temporary migrants living in hostels and more permanent township dwellers are the product of the contradictory injunctions of the colonial economy. The problem with this classical issue in African history is to not overestimate the capacity of colonial control nor to underestimate the socio-spatial effects of this enduring obsession.

Two processes emerged simultaneously during the first half of the twentieth century: first, limited colonial urban governmentality based on the state's identification of a certain number of individualised subjects and, second, a political, bureaucratic and native authority power that produced categories to govern new collective units. As suggested by Fred Cooper, 'if Foucault saw power as "capillary" it was arguably arterial in most colonial contexts –strong near the nodal points of colonial authority, less able to impose its discursive grid elsewhere' (Cooper 2005, pp. 48–49). In 1960, colonial and apartheid power was more arterial in the metropolises of Nigeria and more capillary in those of South Africa. Unsurprisingly, more pervasive state capacity in South Africa meant that specific instruments (passes, reference books, employment offices, municipal censuses) were used to achieve the authorities' panoptic aims towards both temporary migrants and township dwellers. This aim was however limited by bureaucratic constraints and failure (Breckenridge 2014, p. 137, 161) and by the persistence and expansion of an uncontrollable irreducible fringes at its peripheries (squatter areas, overcrowded courtyards in townships). At the same time, we should not underestimate the importance of these classification instruments, which provided various administrations with individualised knowledge pertaining to certain urban categories of the population: township dwellers who enjoyed residential rights, migrants with employment contracts living in hostels with a bedhold number, children in need of care and white juvenile delinquents classified by social services according to their IQs and a few African and coloured delinquents placed in specialised educational institutions. Such devices never existed in Nigeria: individuals were identified through a few instruments and tools (land registers, land use plans, residence certificates) restricted to small-scale portions of urban space (European neighbourhoods, *sabon gari*) and very limited groups (certain heads of households, prostitutes to be sent back home, juvenile delinquents placed in approved schools). Here as in many other colonial contexts, colonial governmentality was incomplete, kept in check by the simultaneous creation of new bureaucratic categories introduced for the purpose of counting, classifying and governing but not individualising. Such apparatus was far to be total in the sense that, despite several attempts introduced by the coercive compassionate late colonial state, there was a large unchecked floating population that could make a living outside colonial surveillance (Balandier 1985; Fair 2018; Fourchard 2001; Gondola 1997). This goes against the idea of an all-pervasive disciplinary colonial power penetrating the smallest details of everyday urban life suggested by a set of urban studies relying more on secondary literature and colonial official narratives than from a detailed confrontation of African and European primary sources.

The manufacturing of invented urban colonial categories has, however, been part of the long-term legacy of urban colonialism, which should not be underestimated either. Moving beyond the literature on resistance to colonialism – often equated with the resistance to capitalism by a number of South African radical historians – helps to reveal the inextricable entanglement between the governing and the governed. The distinction between urbans and migrants resulting from differentiated management of workers institutionalised by the apartheid regime was appropriated by the populations themselves and used for their own self-definition as well as to describe and stigmatise supposedly opposing groups (urbans, semi-urbans, migrants). This official and legal divisions between migrants and township dwellers had enduring legacies long after the end of apartheid. From an urbanist perspective, resurgence of xenophobia in post-apartheid South Africa might also be considered as a combination of this legacy: racial labelling of international migrants partly borrowing from the stereotypes once used by township residents against migrants from reserves; priority for housing requests by township residents based on urban residence rights; squatter mobilisations against foreigners as a legacy of apartheid housing policy freezing urbanisation and a post-apartheid citizenship that prioritises long-standing township residents and their contribution to the collective struggle against new arrivals (Fourchard and Segatti 2015; Landau 2011; Monson 2015; Neocosmos 2010; Nieftagodien 2011). This history that does not coincide with the national narrative regarding the end of apartheid is yet to be written.

Similarly the *indigene/non-indigene* opposition or what is referred to as indigeneity in Nigeria is embedded in a threefold historical process including colonial manufacturing of native/non-native categories, the post-war civil consensus founded on a quota policy giving preference to people that originated in states and local governments, and the return of partisan competition, which has made indigeneity a repertoire capable of mobilising the masses since the return to a civilian regime in 1999. Though researchers and intellectuals in Nigeria have insisted on the destructive power of such a device, neither the federal political elites nor the commissions of inquiry appointed after massacres have fundamentally questioned the institutional arrangements that underpin the equilibrium between federal institutions. These arrangements are defended by state political leaders who see them as a way of eliminating potential candidates from electoral competition. They have also shaped the imaginary of citizens who demand certificates even when they are not necessary and of local employees who consider it 'normal' to exclude *non-indigenes* from the local government. This institutional dimension plays a significant role in naturalising such categories and reveals the power of bureaucratic domination over individuals as much as their willingness to participate to such apparatus.

The urban dimension appears, however, more decisive precisely because the claim over ownership in a locality is central in delineating the boundaries between *indigenes* to include and *non-indigenes* to exclude. Colonial indirect rule reified ethnic boundaries in Nigerian cities but the past 40 years of politicking has been decisive in fuelling ethnoreligious conflicts in the case of Jos and making indigeneity a poorly contested consensus in Ibadan. The genealogy of indigeneity indicates, however, how history matters to legitimise ancestral urban claims. In Ibadan, the dominant narratives are partly located in the nineteenth-century Yoruba urbanism partly reconstructed by post-colonial narratives while in Jos, the origin of the conflict lays in the very colonial foundation of the city. These examples testify to the radical differences in what the term indigeneity means and represents within different administrations and the urban world. Articulating this politics of belonging and this claim to urban ancestral claims is of radical importance to understand why indigeneity is a such an explosive issue in some places while the populations of many other cities or states in the country are indifferent to these forms of discrimination.

It is difficult not to see the degree to which indigeneity exclusion in Nigeria and xenophobic reactions in South Africa overlap with the routine operations of vigilante organisations even if the later could not be reduced to this dimension. The genealogy of police in low-income urban areas of South African and Nigeria actually may help to understand differently the act of urban policing in the global south. There is first a need to highlight the weak presence of colonial police in low-income urban areas and their imperative needs to foster collusive alliances with local authorities. Vigilant groups emerged within this demand of police by the African urban population to deal with property crime and interpersonal violence and the unwillingness of colonial authorities to answer that demand. Anthropology of vigilantism was secondly central to understand that vigilante groups came to exercise a public authority and perform acts of violence against specific threats. New forms of contemporary vigilantism seem thirdly to concur with the urbanisation of neoliberalism theory and its insistence on socio-spatial divisions between an elite able to pay for its own security and subalterns relying on cheap forms of policing such as vigilantism, community policing, lynching and street justice. As a matter of facts, contemporary policing in low-income neighbourhoods in Nigeria and South Africa witnesses the increasing role of not-paid or poorly paid workers providing public goods some of which consist of exercising everyday surveillance under the control of the police while modest households are progressively willing or don't have the choice but to pay for their individual security.

Looking at policing in poor urban areas through the analytical lenses of the urbanisation of neoliberalism or weak state analysis have the inconvenience to focus on what is lacking, either a disappearing social welfare and protective state that never existed in Nigeria nor South Africa (except in its racialised form in South Africa (Seekings and Natrass 2006). This analysis departs from these existing theories to describe what concretely exists on ground. Vigilante groups are the product of a specific urban environment revealing, on the one hand, historical forms of public authority inventing policing practices, identifying specific threats and implementing spatial arrangements relatively stable over time and, on the other hand, more recent vigilante organisations and individuals making a living from security and resorting to a not systematic use of violence.

In rapidly growing low-income neighbourhoods, the objectives of vigilantes and their autonomous action led to manufacturing moral communities, which violently delimited the boundary separating insiders from outsiders. The use of violence was inherent in the routine of these groups and tacitly authorised and sometimes even encouraged by colonial and apartheid administrations. The state attempted to regulate the excessive uses of violence only recently (1987 in Nigeria, 1994 in South Africa to set a nationwide date), but the local arrangements count the most. In the coloured townships of the Cape Flats, the relationship to punishment as well as to authority has radically changed. Corporal punishment, now officially banned, has become less common, more discreet and more invisible, even if it is still seen as a possibility. At the same time, the members of these organisations comply more readily with what the state and the law authorises them to do. In other words, policing of subalterns by other subalterns is not necessarily based on violent acts due to the increasing social and spatial inequalities produced by the neoliberal order. This echoes other research that shows more ambivalent forms of contemporary urban vigilantism, which could pacify the communities and using violence against its perceived enemies, which adopt human rights in practices but reject them in specific instances, which exercise public authority in providing illegal but legitimate forms of punishment not recognised by the state, or which politicised vigilantism through state repression (Smith 2019; Super 2016, 2020; Yonucu 2018).

What is referred to as the commodification of security is one of the other most recent transformations of urban vigilantism, a subject that has seldom been studied until now, except from the perspective of neoliberal privatisation. It is central here again to study historically the dynamics of these predominantly volunteer organisations, which have recently had to cope with a commodification that has opened up job opportunities in areas where the unemployment rate (South Africa) and the self-employment rate (Nigeria) are extremely high. The views of ordinary actors concerned by state sponsored programme do not fit with existing theory: to qualify it as neoliberal misses the fact that this work was previously performed not by police civil servants but by unpaid volunteer groups. As such the programme became popular among unemployed women in low-income Cape Flats as it gave them an opportunity to make a living from this activity. Similarly, in Ibadan and Lagos, the inability to recruit young 'volunteers' indicates essentially a decline in the authority of some elders and their failure to provide effective protection. This inability has opened up (poorly) paid job opportunities to either broader organisations (OPC, Bakassi Boys, Oshodi One Organisations) or individual guards paid by community taxes. In other words, the transformation of urban security seen as a 'community good' into a 'commodified good' reveal essentially a change in the nature of the exercise of public authority in low-income neighbourhoods. The emergence of these forms of public authority and twilight institutions are common in countries at war or in countries in transition (Hoffman and Kirk 2013). The genealogical analysis of vigilantism shows that this exercise is also a more permanent urban condition in low-income neighbourhoods of South Africa and Nigeria.

Despite the tremendous changes in the historical centre of Ibadan and in the Cape Flats in Cape Town, it is actually surprising that threats identified by vigilante groups, like specific social-spatial practices of policing have hardly changed over time. By identifying specific targets – township youth and migrants from hostels in the case of Cape Town and outsiders in the neighbourhood who are unfamiliar with its regulations in the case of Ibadan – these have played a key role in constructing the archetypal figures of exclusion, which are not fundamentally different from those produced by colonial engineering or apartheid identified in the first part of the book. This has concrete spatial and temporal implications that have not so much changed over time. This is very clear in the regulation of behaviour on weekends and by night. Each night or every weekend groups reactivate the boundary line between residents and non-residents within the perimeter of the neighbourhood, as the patrols have neither a reason nor permission to venture outside them. But the relationship to nocturnal danger in Nigeria has been constructed differently than in South Africa, mainly because the threat is identified as 'different'. The adoption of a citywide curfew in Lagos or Ibadan resulted in shifting social activities to daytime hours and withdrawing into family space at night, a change amply attested by the extraordinary popularity of Nollywood home videos in the 1990s that could be watched without leaving the comfort of the family's compound or of the local neighbourhood (Jedlowski 2012, p. 438). Even if the night-life economy has been back in Lagos in the past fifteen years – it remains highly concentrated in some specific areas – most low-income neighbourhoods are, like residential middle-class or upper-class neighbourhoods, often closed by night or under strict surveillance as the threat is always perceived as coming from outside the neighbourhood. In Cape Town, on the other hand, the imposition of a curfew in the Cape Flats never led to such nocturnal anomies precisely because it was exclusively aimed at township youth and children who were supposed to stay home. The weekly surveillance of moral behaviour by parents and grandparents at night continues to be a standard part of these organisations as before. Belonging to the world of adults remains a defining feature of these policing groups and youth their main targets, a continuity that the struggle of young anti-apartheid activists in the 1980s ultimately never called into question. Even now, the adult volunteers perceive children and youth as responsible for social disorder in the life of the township and link them to a number of activities that are legal nowadays but still considered suspect at night (loitering in the streets, drinking at the *shebeen*, visiting prostitutes, going out unaccompanied by an adult). This is the remaining part of this forgotten struggle of vigilante groups operating during colonial and apartheid periods. While the new democratic regime did not eradicate these representations or certain former practices of social control such as body searches, the relationship between young residents and those exercising this adult authority is changing. Although young people submit to control procedures that would be unimaginable in other contexts, they are no longer reluctant to defy the organisations within legally permissible limits – practices that have yet to be explored in the literature.

Moving for the police in and of the neighbourhood in Nigeria and South Africa to motor parks and local government offices in Lagos, Ibadan and Jos is another way of understanding urban politics from concrete places of power. Instead of starting from a definition of urban politics as a more or less radical contestation of the state, this ethnographic journey of urban daily life explores confrontation, cooperation and compromise between politicians, bureaucrats, union leaders and residents, their uneven politicisation and the elusive boundaries between state and non-state actors in metropolises dominated by labour precarity. The wish to participate to power, the daily work of street-level bureaucrats and those acting on their behalf, the appropriation and mundane uses of bureaucratic and patronage domination by subaltern actors reveal a new urban politics at work. Following such an approach allows to reassess the radical importance of vertical ties within networks of 'informal' actors, which can bring opportunities for political inclusion, help to extend service deliveries or lead to violent mobilisations or exclude entire groups to the access of state resources.

All these practices, which make up the daily life of millions of actors, essentially revolve around negotiating a share of urban taxation. Adding urban revenue to federal revenue allows us to see the bigger picture. It is the combination of these two resources from the street and from the office that is pivotal in implementing public action. For obvious reasons the Metropolitan project of Lagos has attracted a great deal of attention in Nigeria. For researchers, for the governors of the other states, for the Nigerian political class and, now, for the international development agencies, Lagos state is more than ever a space for experimenting and implementing public policies and financial autonomy based on the relative consent of the population to pay income tax. This is one side of the Lagos success story however coming mainly from policy and officials' circles.

These innovations should not make us forget the processes that reproduce predatory mechanisms of the urban economy. The government of transport in Lagos, Ibadan and many other Nigerian cities has been left to the discretion of the NURTW, which has free hands to raise revenue from the daily transport of millions of Nigerians. Like vigilante groups providing security in poor neighbourhoods, the union exercises an almost exclusive public authority over the urban transport system in the country. By being located between the state and the urban economy, this institution has become so powerful that it goes beyond the 'quiet encroachment of the ordinary' and beyond 'the state as an informalised entity' a concept which does not sufficiently identify the uneven bargaining power of the most important political and economic actors. If the informal and the illegal are constitutive parts of urban politics in most cities of the world (Boudreau 2017) the key issue is to explore the ways in which actors cross the frontier between the legal and the illegal. The union is performing public authority like any state bureaucracy and union members act as street-level bureaucrats. Their daily work based in motor parks and the streets is fundamentally based on a suspension of the law, which is of benefit to official networks of elected politicians, bureaucrats and the police. These illegal arrangements produced a form of urban order which is regularly challenged by episodes of violence between union members. The numerous victims of faction fights and political violence are the dark side of this politicised government of motor parks. Key union leaders loyal to the governor in power have never been charged before court despite the number of murders involving their members. The power of the union is based on this capacity to use both this extraordinary violence in the street and to be routinely located between the most powerful patronage networks and formal institutions in the city. The politicisation of motor parks is situated at this crossroads. While the political economy of motor parks cannot be fully understood without looking at the 'office' side – its official dimension – bureaucratic documents produced within local governments are constitutive of forms of urban sociality and conflicts as suggested by Matthew Hull (2012a). The comparative analysis clearly indicates that the production of the same documents is at the heart of most of the political battles in Jos while it is a marginal matter in Ibadan. The situation in Ibadan is perhaps closer to what is happening in most parts of Nigeria: an apparatus that turns *non-indigenes* into second-class citizens ultimately generates little or no protest. Despite the official existence of such a discriminatory official and bureaucratic apparatus, nobody is asking for its suppression or even a temporary suspension even if it fuels tensions and violent conflicts among groups. The certificates produced in local government offices like the dues collected in motor parks show that fundamental official and non-official forms of extortion and discrimination with their incredible potential for violence have become so entrenched in Nigerian urban society than no political or bureaucratic power is willing to get rid of.

References

Abrahams, R. (2007). 'Some thoughts on the comparative study of vigilantism'. In: *Global Vigilantes: Perspectives on Justice and Violence* (eds. D. Pratten and A. Sen). London, Hurst.

Abrahamsen, R. and Williams, M. (2009). 'Security beyond the State: Global security assemblages in international politics'. *International Political Sociology*, 3, 1–17.

Abu-Lughod, J. (1980). *Rabat. Urban Apartheid in Morocco*. Princeton, Princeton University Press.

Adeboye, O. (2003). 'Intra-ethnic segregation in colonial Ibadan: The case of Ijebu settlers'. In: *Security Crime and Segregation in West African Cities since the 19th Century* (ed. L. Fourchard and I.O. Albert), 303–320. Ibadan IFRA, Paris Karthala.

Adeboye, O. (2007). '"Iku Ya J'esin" politically motivated suicide, social honor and chieftaincy politics in early colonial Ibadan'. *Canadian Journal of African Studies*, 41 (2), 189–225.

Adebanwi, W. (2009). 'Terror, territoriality and the struggle for indigeneity and citizenship in Northern Nigeria'. *Citizenship Studies*, 13 (4), 349–363.

Adebanwi, W. and Obadare, E. (eds.) (2013). *Democracy and Prebendalism in Nigeria. Critical Interpretations*. New York, Palgrave Macmillan.

Adebanwi, W. (2014). *Yoruba Elites and Ethnic Politics in Nigeria. Obafemi Awolowo and Corporate Agency*. Cambridge, Cambridge University Press.

Adenugba, A.A. and Ogechi, C.F. (2013). 'The effects of internal revenue generation on infrastructural development. A study of Lagos State Internal Revenue Service'. *Journal of Educational and Social Research*, 3 (2), 419–436.

Aderinto, S. (2012). 'The problem of Nigeria is slavery, not white traffic, globalisation and the politicisation of prostitution in Southern Nigeria, 1921-1955'. *Canadian Journal of African Studies*, 46 (1), 1–22.

Aderinto, S. (2014). *When Sex Threatened the State: Illicit Sexuality, Nationalism and Politics in Colonial Nigeria, 1900-1958*. Urbana Chicago, University of Illinois Press.

Aderinto, S. (2018). *Guns and Society in Colonial Nigeria. Firearms, Culture and Public Order*. Bloomington, Indiana University Press.

Agbaje, A. (2003). 'Personal rule and regional politics: Ibadan under military regimes, 1986-1996'. In: *Money Struggles and City Life Devaluation in Ibadan and Other Urban Centers in Southern Nigeria 1986-1996* (eds. J.I. Guyer, L. Denzer and A. Agbaje). Ibadan, Bookbuilders.

Agbiboa, D.E. (2018). 'Informal urban governance and predatory politics in Africa: The role of motor-park touts in Lagos'. *African Affairs*, 117, 466, 62–82.

Agbiboa, D.E. (2020). 'Between cooperation and conflict: The national union of road transport workers in Lagos, Nigeria'. *Crime, Law and Social Change*, early view.

Agbola, T. (1997). *Architecture of Fear: Urban Design and Construction Response to Urban Violence in Lagos, Nigeria*. Ibadan, IFRA.

Agbu, O. (2004). 'Ethnic militias and the threat to democracy in post transition Nigeria'. Working paper 127. Uppsala, Nordiska Afrikainstitutet.

Agrikoliansky, É. (2001). 'Carrières militantes, et vocation à la morale: Les militants de la Ligue des droits de l'homme dans les années 1980'. *Revue française de science politique*, 51 (1), 27–46.

Akerele, W. O. (1997). *The Effects of Economic Adjustment on Employment in the Urban Informal Sector of Ibadan City*. Niser, Ibadan.

Ahire, P.T. (1991). *Imperial Policing. The Emergence and Role of the Police in Colonial Nigeria, 1860–1960*. Buckingham, Open University Press.

Aiyede, R. (2003). 'The dynamics of civil society and the democratization process in Nigeria'. *Canadian Journal of African Studies*, 37 (1), 1–27.

Ajala, A.S. (2008). 'Identity and space in Ibadan Politics, Western Nigeria'. *African Identities*, 6 (2), 149–168.

Akinyele, R.T. (2007). 'The involvement of the Oodua People's Congress in crime control in South-Western Nigerian cities'. In: *Gouverner les villes d'Afrique. État, gouvernement local et acteurs privés* (dir. L. Fourchard), 139–161. Paris, Karthala.

Akinyele, R.T. (2009). 'Contesting for space in an urban centre: The Omo Onile Syndrome in Lagos'. In: *African Cities. Competing Claims on Urban Spaces* (eds. F. Locatelli and P. Nugent), 55–81. Leyde, Brill.

Albert, I.O. (1994). 'Violence in Metropolitan Kano: A historical perspective'. In: *Urban Violence in Africa* (eds. E. Osaghae et al.), 111–138. Ibadan, IFRA.

Albert, I.O. (1996). 'Ethnic residential segregation in Kano, Nigeria and its Antecedents'. *African Study Monographs*, 17 (2), 85–100.

Albert, I.O. (2007). 'Between the State and transport unions: NURTW and the politics of managing public motor parks in Ibadan and Lagos, Nigeria'. In: *Gouverner les villes d'Afrique: État, gouvernement local et acteurs privés* (dir. L. Fourchard). Paris, Karthala.

Alexander, P.F. (2009). *Alan Paton, Selected Letters*. Le Cap, Van Riebeeck Society.

Alexander, J. and Kynoch, G. (2011). 'History and legacy of punishment in Southern Africa'. *Journal of Southern African Studies*, 37 (3), 398–403.

Amselle, J.-L. and M'bokolo, E. (eds.) (1985). *Au cœur de l'ethnie. Ethnies, tribalisme et État en Afrique*. Paris, La Découverte.

Anderson, D.B. (2005). *Histories of the Hanged, Britain's Dirty War on Kenya and the End of Empire*. Londres, Phoenix.

Anderson, D.B. and Killingray, D. (eds.) (1991). *Policing the Empire. Government, Authority and Control, 1830–1940*. Manchester, Manchester University Press.

Anderson, D.B. and Killingray, D. (1992). 'An orderly retreat? Policing the end of empire'. In: *Policing and Decolonisation. Politics, Nationalism and the Police,*

1917–1965 (eds. D.B. Anderson and D. Killingray), 1–21. Manchester, Manchester University Press.

Anifowose, R. (1982). *Violence and Politics in Nigeria. The TIV and Yoruba Experience.* Enugu, Nok.

Animasaiwun, G.A. (2013). 'Godfathering in Nigeria's Fourth Republic: The pyramid of violence and political insecurity in Ibadan, Oyo State, Nigeria'. IFRA Nigeria e-papers Series, 27.

Annual Report of the Nigeria Police Force, 1940.

Anthony, D.D. (2002). *Poison and Medicine: Ethnicity, Power, and Violence in a Nigerian City, 1966 to 1986.* Oxford, James Currey.

Arnold, D. (1986). *Police Power and Colonial Rule: Madras 1859–1947.* Oxford, Oxford University Press.

Auyero, J. (2000). 'The logic of clientelism in Argentina: An ethnographic account'. *Latin American Research Review*, 35 (3), 55–81.

Awasom, N.F. (2003). Hausa Traders, Residential Segregation and the Quest for Security in 20th Century Colonial Bamenda Township (Cameroon). In: *Sécurité, crime et ségrégation dans les villes d'Afrique de l'Ouest du XIXᵉ siècle à nos jours* (dir. L. Fourchard and I.O. Albert). Paris, Karthala/ Ibadan, IFRA.

Awenengo Dalberto, S. and Banegas, R. (2018). 'Citoyens de papier: Des écritures bureaucratiques de soi en Afrique'. *Genèses*, 112, 3.

Bach, D.C. (2006). 'Inching towards a country without a state: Prebendalism, violence and state betrayal in Nigeria'. In *Big African States* (eds. C. Clapham, J. Herbst and G. Mills), 63–96. Johannesbourg, Wits University Press.

Badenhorst, C. and Mather, C. (1997). 'Tribal recreation and recreating tribalism: Culture, leisure and social control on South Africa's Gold Mines, 1940–1950'. *Journal of Southern African Studies*, 23 (3), 473–489.

Badroodien, A. (2001). *A History of the Ottery School of Industries in Cape Town. Issues of Race, Welfare and Social Order in the Period 1937 to 1968*, Ph D in History, Cape Town, University of the Western Cape.

Bähre, E. (2014). 'A trickle-up economy: Mutuality, freedom and violence in Cape Town's taxi associations'. *Africa*, 84 (4), 576–594.

Bailleau, F., Cartuyvels, Y. and De Fraene, D. (2009). 'La justice pénale des mineurs en Europe et ses évolutions. La criminalisation des mineurs et le jeu des sanctions'. *Déviance et société*, 33 (3), 255–269.

Baker, B. (2008). *Multi-Choice Policing in Africa.* Uppsala, Nordiska Afrikainstitutet.

Baines, G. (1989). 'The control and administration of Port Elizabeth's African population, c. 1834–1923'. In: *Contree / [Raad vir Geesteswetenskaplike Navorsing, Instituut vir Geskiedenisnavorsing, Afdeling Streekgeskiedenis]*.

Baines, G. (1994). 'The contradiction of community politics. The African petty bourgeoisie and the new Brighton advisory Board, 1937–1952'. *Journal of African History*, 35, 79–97.

Balandier, G. (1985). *Sociologie des Brazzavilles noires.* Paris, Presses de Sciences Po, [1955.] 2ᵉ éd.

Balbo, M. (1993). 'Urban planning and the fragmented city of developing countries'. *Third World Planning Review*, 15 (1), 23–36.

Banégas, R. (1998). 'Marchandisation du vote, citoyenneté et consolidation démocratique au Bénin'. *Politique africaine*, 69, 78–79.

Banégas, R. (2006). 'Côte d'Ivoire: Patriotism, ethnonationalism and other African modes of self-writing'. *African Affairs*, 105 (421), 535–552.

Barber, K. (1991). *I Could Speak Until Tomorrow: Oriki, Women and the Past in a Yoruba Town*, 203. Londres, International African Institute.

Barchiesi, F. (2019). 'Precarious and informal labour'. In: *General Labour History of Africa. Workers, Employers and Governments* (eds. S. Belluci and A. Eckert), 45–76. James Currey, Suffolk, Boydell and Brewer, Rochester, International Labour Organisation, Abidjan.

Barkindo, B.M. (1983). *Studies in the History of Kano*. Ibadan, Heinemann Educational Books.

Barnes, S. (1986). *Patrons and Power: Creating a Political Community in Metropolitan Lagos*. Manchester, Manchester University Press.

Barth, H. (1857). *Travels and Discoveries in North and Central Africa: Being a Journal of an Expedition undertaken under the Auspices of H.B.M.'s Government, in the Years 1849–1855*. New York; (NY), Harper and Brothers.

Basorun, J.O. and Rotowa, O.O. (2012). 'Regional assessment of public transport operations in Nigerian cities, the case of Lagos Island'. *International Journal of Developing Society*, 1 (2), 82–87.

Bat, J.-P. (2012). *Le Syndrome Foccart. La politique africaine en Afrique de 1959 à nos jours*. Paris, Gallimard.

Bayart, J.-F. (2008a). 'Comparer en France. Petit essai d'autobiographie disciplinaire'. *Politix*, 3 (83), 205–232.

Bayart, J.-F. (2008b). 'Hégémonie et coercition en Afrique subsaharienne. La "politique de la chicotte"'. *Politique africaine*, 110, 123–152.

Bayart, J.-F. (1989). *L'État en Afrique. La politique du ventre*. Paris, Fayard.

Bayart, J.-F. (2007). 'Les chemins de traverse de l'hégémonie coloniale en Afrique de l'Ouest francophone. Anciens esclaves, anciens combattants, nouveaux musulmans'. *Politique africaine*, 105 (1), 201–240.

Bayart, J.-F. and Bertrand, R. (2006). 'De quel legs colonial parle-t-on?'. *Esprit*, 12.

Bayart, J.-F., Geschiere, P., and Nyamnjoh, F. (2001). 'Autochtonie, démocratie, citoyenneté en Afrique'. *Critique internationale*, 10, 177–194.

Bayart, J.-F., Mbembe, A. and Toulabor, C. (1992). *Le Politique par le bas. Contributions à une problématique de la démocratie*, Paris, Karthala.

Bayat, A. (1997). *Street politics, Poor People's Movements in Iran*. New York; (NY), Columbia University Press.

Bayat, A. (2000). 'From "dangerous classes" to "quiet rebels". Politics of the urban subaltern in the global south'. *International Sociology*, 15 (3), 533–557.

Bayly, C.A., Beckert, S., Connelly, M., Hofmeyr, I., Kozol, W. and Seed, P. (2006). 'American historical review conversation: On transnational history'. *American Historical Review*, 111 (5), 1441–1464.

Beinart, W. (1992). 'Political and collective violence in Southern African Historiography'. *Journal of Southern African Studies*, 18 (3), 455–486.

Bekker, S. and Fourchard, L. (eds.) (2013). *Governing Cities in Africa, Politics and Policies*. Pretoria, HSRC Press.

Belina, B. and Helms, G. (2003). 'Zero tolerance for the industrial past and other threats: Policing and urban entrepreneurialism in Britain and Germany'. *Urban Studies*, 40 (9), 1845–1867.

Bénit-Gbaffou, C. (2016). 'Do street traders have the "right to the city"? The politics of street trader organizations in inner city Johannesburg, post-Operation Clean Sweep'. *Third World Quarterly*, 37 (6), 1102–1129.

Bénit-Gbaffou, C., Fourchard, L. and Wafer, A. (2012). 'Local politics and the circulation of community security initiatives in Johannesburg'. *International Journal of Urban and Regional Research*, 36 (5), 936–957.

Bénit-Gbaffou, C. et al. (2013). 'Exploring the role of party politics in the governance of African cities'. In: *Governing Cities in Africa, Politics and Policies* (eds. S. Bekker and L. Fourchard). Pretoria, HSRC Press.

Bénit-Gbaffou, C., Owuor, S. and Fabiyi, S. (2011). 'The impact of enclosed neighborhoods on privatization of public space: A comparative analysis of Nairobi, Johannesburg and Ibadan'. *Regional Development Studies*, 15, 72–85.

Berg, J., Akinyele, R., Fourchard, L., van der Waal, K. and Williams, M. (2013). 'Contested social orders. Negotiating urban security in Nigeria and South Africa'. In: *Governing Cities in Africa, Politics and Policies* (eds. S. Bekker and L. Fourchard). Pretoria, HSRC Press.

Berg, J. and Shearing, C. (2015). 'New authorities: Relating state and non-state security auspices in South African improvement districts'. In: *Policing and the Politics of Order-Making* (eds. H. Kyed, and P. Albrecht). Oxon, Routledge.

Berman, B. and Lonsdale, J. (1992). *Unhappy Valley. Conflict in Kenya and Africa. State and Class*, vol. 1. Oxford, James Currey/Athens (Ohio), Ohio University Press.

Bernault, F. (dir.) (1999). *Enfermement, prison et châtiments en Afrique. Du XIXᵉ siècle à nos jours*. Paris, Karthala.

Bertrand, R. (2006a). 'Les sciences sociales et le moment colonial: de la problématique de la domination coloniale à celle de l'hégémonie coloniale'. FNSP/ CERI, *Questions de recherche*, 18.

Bertrand, R. (2006b). Mémoires d'empire. *La controverse autour du 'fait colonial'*. Bellecombe-en-Bauges, Éditions du croquant.

Bickford-Smith, V. (2008). 'Urban history in the New South Africa: Continuity and innovation since the end of Apartheid'. *Urban History*, 35, 288–315.

Bierschenk, T. and Olivier de Sardan, J.-P. (eds.) (2014). *States at Work: Dynamics of African Bureaucracies*, 42. Leyde, Brill.

Bigon, L. (2016). 'Bubonic plague, colonial ideologies, and urban planning policies: Dakar, Lagos, and Kumasi'. *Planning Perspectives*, 31 (2), 205–226.

Bissell, W.C. (2011). *Urban Design, Chaos and Colonial Power in Zanzibar*. Bloomington (IN), Indiana University Press.

Blanchard, E. (2012). 'Ordre colonial'. *Genèses*, 86 (1), 2–7.

Blanchard, E., Deluermoz, Q. and Glasman, J. (2011). 'La professionnalisation policière en situation coloniale, détour conceptuel et explorations historiographiques'. *Crime, Histoire et Sociétés*, 15 (2), 3–53.

Blanchard, E. and Glasman, J. (2012). 'Le maintien de l'ordre dans l'Empire français: Une historiographie émergente'. In: *Maintenir l'ordre colonial* (dir. J.-P. Bat and N. Courtin), 11–13. Rennes, Presses universitaires de Rennes.

Bloembergen, M. (2012). 'Vol meurtre et action policière dans les villages javanais'. *Genèses*, 86 (1), 8–36.

Blundo, G. and Olivier De Sardan, J.-P. (dir.) (2007). *État et corruption en Afrique. Une anthropologie comparative des relations entre fonctionnaires et usagers (Bénin, Niger, Sénégal)*. Paris, Karthala.

Bøås, M. and Dunn, K. (2013). *Politics of Origin in Africa: Autochthony, Citizenship and Conflict*. Pretoria, Human Sciences Research Council Press.

Bodea, C. and Lebas, A. (2016). 'The origins of voluntary compliance: Attitudes toward taxation in urban Nigeria'. *British Journal of Political Science*, 46 (1), 215–238.

Bonnecase, V. (2011). *La Pauvreté au Sahel. Du savoir colonial à la mesure internationale*. Paris, Karthala.

Bonner, P. and Nieftagodien, N. (2008). *Alexandra: A History*. Johannesburg, Wits University Press.

Bonnet, F. (2019). *The Upper Limit: How Low Wage Work Defines Punishment and Welfare*, Oakland, University of California Press.

Botiveau, R. (2014). 'The politics of Marikana and South Africa's changing labour relations'. *African Affairs*, 113 (450), 128–137.

Boudreau, J.A. (2017). *Global Urban Politics. Informalization of the State*. Maden, Polity Press.

Bourdelais, P. (2005). 'L'intolérable du travail des enfants: Son émergence et son évolution entre compassion et libéralisme en Angleterre et en France'. In: D *Les Constructions de l'intolérable* (dir. Fassin and P. Bourdelais). Paris, La Découverte.

Bourdieu, P. (2000). *Propos sur le champ politique*. Lyon, Presses Universitaires de Lyon.

Bozzoli, B. (1987). 'Class, community and ideology in the evolution of South African Society'. In: *Class, Community and Conflict. South African Perspectives* (ed. B. Bozzoli), 1–43. Johannesburg, Ravan Press.

Branch, D. (2009). *Defeating Mau Mau, Creating Kenya, Counterinsurgency, Civil War and Decolonization*. Cambridge, Cambridge University Press.

Branche, R. (2001). *La torture et l'armée pendant la guerre d'Algérie (1954–1962)*. Paris, Gallimard.

Breckenridge, K. (2005). 'Verwoerd's bureau of proof: Total information in the making of apartheid'. *History Workshop Journal*, 59 (1), 83–108.

Breckenridge, K. (2014). *Biometric State. The Global Politics of Identification and Surveillance in South Africa, 1850 to the Present*. Cambridge, Cambridge University Press.

Breckenridge, K. and Szreter, S. (eds.) (2012). *Registration and Recognition: Documenting the Person in World History*. Oxford, Oxford University Press.

Brenner, N. and Theodore, N. (2002). 'Cities and the geographies of existing neoliberalism'. *Antipode*, 34 (3), 349–379.

Brewer, J. (1994). *Black and Blue. Policing in South Africa*. Oxford, Clarendon Press.

Briquet, J.-L. and Sawicki, F. (dir.) (1998). *Le Clientélisme politique dans les sociétés contemporaines*. Paris, PUF.

Brogden, M. and Nijhar, P. (2013). *Community Policing. National and International Models and Approaches*. London and New York;, Routledge.

Brown, C. (2003). *We Were All Slaves, African Miners, Culture and Resistance at the Enugu Government Colliery*. Oxford, James Currey/Le Cap, David Philip/ Portsmouth (NH), Heinemann.

Brown, A. (2004). 'Mythologies and panics, XXth century constructions of child prostitution'. *Children and Society*, 18, 344–354.

Buire, C. (2019). *Citadins-citoyens au Cap. Justice et espace après l'apartheid*. Ifas, Johannesburg, Presses Universitaires de Paris Nanterre.

Burgess, E.W. (1926). *The Urban Community*. Chicago (IL), University of Chicago Press.

Burton, A. (2001). 'Urchins, loafers and the cult of the cowboys: Urbanisation and delinquency in Dar es Salaam, 1919–1961'. *Journal of African History*, 42 (1), 200.

Burton, A. (2005). *African Underclass: Urbanisation, crime and colonial order in Dar es Salaam*. Oxford: James Currey.

Burton, A. (2007). 'The Haven of Peace purged: Tackling the undesirable and unproductive poor in Dar es Salaam, ca. 1950s–1980s'. *The International Journal of African Historical Studies*, 40 (1), 119–151.

Buur, L. (2005). 'The sovereign outsourced: Local justice and violence in Port Elizabeth'. In: *Sovereign Bodies. Citizens, Migrants and States in the Postcolonial World* (eds. T. Blom Hansen and F. Stepputat), 192–218. Princeton (NJ), Princeton University Press.

Buur, L. (2006). 'Reordering society: Vigilantism and expressions of sovereignty in Port Elizabeth's Townships'. *Development and Change*, 37 (4), 735–757.

Buur, L. and Jensen, S. (2004). 'Introduction: Vigilantism and the policing of everyday life in South Africa'. *African Studies*, 63 (2), 139–152.

Cabane, L. (2012). 'Gouverner les catastrophes. Politiques, savoirs et organisation de la gestion des catastrophes en Afrique du Sud'. Ph D in Political Science, Paris, Sciences Po.

Caldeira, T. (2000). *City of Walls, Crime, Segregation and Citizenship in Sao Paulo*. Berkeley (CA), University of California Press.

Campbell, C. (2006). 'Juvenile Delinquency in Colonial Kenya, 1900–1939'. *The Historical Journal*, 45 (1), 142–143.

Carrel, M. and Neveu, C. (2014). *Citoyennetés ordinaires. Pour une approche renouvelée des pratiques citoyennes*. Paris, Karthala.

Carrier, N. (2010). 'Sociologies anglo-saxonnes du virage punitif. Timidité critique, perspectives totalisantes et réductrices'. *Champ Pénal, Penal Field*, 7, on line.

Celik, Z. (1997). *Urban Forms and Colonial Confrontations: Algiers under French Rule*. Berkeley (CA), University of California Press.

Chalfin, B. (2010). *Neoliberal Frontiers: An Ethnography of Sovereignty in West Africa*. Chicago (IL), University of Chicago Press.

Chandavarkar, R. (1998). *Imperial Power and Popular Politics: Class, Resistance and the State in India, c. 1850–1940*, Cambridge, Cambridge University Press.

Chaskalon, M. (1986). 'The road to Sharpeville'. *African Studies Seminar Paper*, 199, Wits University, African Studies Institute.

Cheeseman, N. and De Gramont, D. (2017). 'Managing a mega-city: Learning the lessons from Lagos'. *Oxford Review of Economic Policy*, 33 (3), 457–477.

Chiranchi, A. Y. (2001). The Native Authority Police and the Maintenance of Law and Order in Kano Emirate, 1925-1968, in Abdalla Uba Adamu, Chieftancy and Security in Nigeria: Past Present and Future, *Proceedings of the National Conference on Chieftaincy and Security in Nigeria to commemorate the 40th anniversary of His Royal Highness*, the Emir of Kano, pp. 256–280.

Chisholm, L. (1991). 'Education, punishment and the contradictions of penal reform: Alan Paton and Diepkloof Reformatory, 1934–1948'. *Journal of Southern African Studies*, 17 (1), 23–42.

Chisholm, L. (n.d.). 'Aspects of child-saving in South Africa. Classifying and segregating the delinquent: The struggle over the reformatory, 1917–1934'. unpublished article.

Choplin, A. and Ciovalella, R. 'Gramsci and the African Città Futura: Urban subaltern politics from the margins of Nouakchott, Mauritania'. *Antipodes*, 2016.

Chrétien, J.P. and Prunier, G. (eds.) (2003). *Les ethnies ont une histoire*. Paris, Karthala.

Cissokho, S. (2016). Le contrat social sénégalais au ras du bitume (1985–2014): De la formation du groupe professionnel des chauffeurs au renforcement des institutions politiques. PhD in political science, University of Panthéon Sorbonne.

Cohen, A. (1969). *Custom and Politics in Urban Africa: A Study of Hausa Migrants in Yoruba Towns*. Berkeley (CA), University of California Press.

Coldham, S. (2000). 'Criminal justice policies in commonwealth Africa: Trends and prospects'. *Journal of African Law*, 44 (2), 218–238.

Cole, J. (1987). *Crossroads: The Politics of Reform and Repression, 1976–1986*. Johannesburg, Ravan Press.

Cook, A. (1982). *Akin to Slavery: Prison Labour in South Africa*. Londres, International Defence and Aid Fund.

Comaroff, J. and Comaroff, J.L. (2006). *Law and Disorder in the Postcolony*. Chicago (IL), University of Chicago Press.

Comaroff, J. and Comaroff, J.L. (2012). *Theory from the South: Or, How Euro-America is Evolving toward Africa*. Boulder (CO), Paradigm Publishers.

Cooper, F. (ed.) (1983). *Struggle for the City: Migrant, Labor, Capital and the State in Urban Africa*. Beverly Hills (CA), Sage Publications.

Cooper, F. (1994). 'Conflict and connection: Rethinking Colonial African history'. *American Historical Review*, 99, 1516–1545.

Cooper, F. (1995). *Decolonization and African Societies. The Labor Question in French and British Africa*. Cambridge, Cambridge University Press.

Cooper, F. (2002). *Africa since 1940s*. Cambridge, Cambridge University Press.

Cooper, F. (2005). *Colonialism in Question: Theory, Knowledge, History*. Berkeley (CA), University of California Press.

Cooper, F. (2012). 'Voting, welfare and registration: The strange fate of the État-Civil in French Africa, 1945–1960'. In: *Registration and Recognition: Documenting the Person in World History* (eds. K. Breckenridge and S. Szreter). Oxford, Oxford University Press.

Coquery-Vidrovitch, C. (1993). *Histoire des villes d'Afrique noire. Des origines à la colonisation*. Paris, Albin Michel.

Corrigall-Brown, C. (2012). *Patterns of Protest. Trajectories of Participation in Social Movements*. Redwood City (CA), Stanford University Press.

Cristelow, A. (2005). 'Land rights, commerce and authority in Kano'. In: *Land, Literacy and the State in Sudanic Africa* (ed. D. Crummey), 256–262. Trenton (NJ)/Asmara, Red Sea Press.

Crush, J., Jeeves, A. and Yudelman, D. (1991). *South Africa's Labor Empire: A History of Black Migrancy to the Gold Mines*. Le Cap, David Philip.

David, B. (2007). 'Good cops? Bad Cops? Assessing the South African Police Service'. *South African Crime Quarterly*, 21.

Davis, M. (1992). *City of Quartz: Excavating the Future in Los Angeles*. New York; (NY), Vintage.

De Boeck, P. and Plissart, M.-F. (2004). *Kinshasa. Tales of the Invisible City*. Tervuren, Royal Museum for Central Africa.

Deltombe, T., Domergue, M. and Tatsitsa, J. (2011). *Kamerun! Une guerre cachée aux origines de la Franceafrique, 1948–1971*. Paris, La Découverte.

Demarest, L., Langer, A. and Ukiwo, U. (2020). 'Nigeria's Federal Character Commission (FCC): A critical appraisal'. *Oxford Development Studies*.

Demeestere, R. (2016). "Même si tu as tes papiers, ils t'embarquent'. La gestion policière de l'immigration africaine ou l'institutionnalisation de la xénophobie dans l'Afrique du Sud post-apartheid'. *Politique africaine*, 142 (2), 145–167.

Dewey, M. (2018). 'Zona liberada: La suspensión de la ley como patrón de comportamiento estatal'. *Nueva Sociedad*, 276, 102–117.

Dikec, M. and Swyngedouw, E. (2017). 'Theorizing the politicising city'. *International Journal of Urban and Regional Research*, 41 (1), 2–18.

Didier, S., Peyroux, É. and Morange, M. (2012). 'The spreading of the city improvement district model in Johannesburg and Cape Town: Urban regeneration and the neoliberal agenda in South Africa'. *International Journal of Urban and Regional Research*, 36 (5), 915–935.

Diphoorn, T. (2016). *Twilight Policing. Private Security and Violence in Urban South Africa*. Oakland, University of California Press.

Dobry, M. (1986). *Sociologie des crises politiques: La dynamique des mobilisations multisectorielles*. Paris, Presses de Sciences Po.

Doris, D.T. (2011). *Vigilant Things: On Thieves, Yoruba Anti-Aesthetics and the Strange Fats of Ordinary Objects in Nigeria*. University of Washington Press

Dorman, S., Hammett, D. and Nugent, P. (eds.) (2007). *Making Nations, Creating Strangers: States and Citizenship in Africa*. Leyde, Brill.

Dubow, S. (2001). 'Scientism, social research and the limits of "South Africanism": The case of Ernst Gideon Malherbe'. *South African Historical Journal*, 44 (1), 99–142

Dubow, S. and Jeeves, A. (eds.) (2005). *South Africa's 1940s: World of Possibilities*. Le Cap, Double Storey Books.

Dubresson, A. and Jaglin, S. (dir.) (2009). *Le Cap après l'Apartheid. Gouvernance métropolitaine et changement urbain*. Paris, Karthala.

Eberlein, R. (2006). 'On the road to the state's perdition? Authority and sovereignty in the Niger Delta, Nigeria'. *Journal of Modern African Studies*, 44 (4), 573–596.

Echenberg, M. (2002). *Black Death, White Medicine: Bubonic Plague and the Politics of Public Health in Colonial Senegal, 1914–1945*. Portsmouth, Heinemann/Oxford, James Currey/Le Cap, David Philip.

Eckert, A. (2006). 'Urbanisation in colonial and postcolonial West Africa'. In: *Themes in West Africa's History* (ed. E.K. Akyeampong), 213. Oxford, James Currey/Athens (Ohio), Ohio University Press/Accra, Woeli.

Eckert, A. (2004). 'Regulating the social: Social security, social welfare and the state in late colonial Tanzania'. *Journal of African History*, 45 (3), 467–489.

Eckert, A. (2019). 'Wage labor'. In: *General Labour History of Africa. Workers, Employers and Governments* (eds. S. Belluci and A. Eckert), 17–44. James Currey, Suffolk, Boydell and Brewer, Rochester, International Labour Organisation, Abidjan.

Ekeh, P. (ed.) (2007). *History of the Urhobo People of Niger Delta*. Ikeja, Urhobo Historical Society.

Eklûf Amirell, S. (2009). 'La piraterie maritime en Afrique contemporaine. Ressorts locaux et internationaux des activités de piraterie au Nigeria et en Somalie'. *Politique africaine*, 116 (4), 97–119.

Ekpo-Otu, M. U. (2013). 'Contestations of identity: Colonial policing of female sexuality in the Cross-river region of Southern Nigeria'. *Inkanyiso, Journal of Humanities and Social Sciences*, 5, 1.

Ellis, S. (2011). 'The genesis of the ANC's armed struggle in South Africa, 1948–1961'. *Journal of Southern African Studies*, 37 (4), 657–676.

Enweremadu, D.U. (2012). Anti-Corruption Campaign in Nigeria (1999–2007), The Politics of a Failed Reform. African studies center, French Institute for Research in Africa, Leiden.

Enwezor, Okwui et al. (eds.), (2002). *Under Siege: Four African Cities. Freetown, Johannesbourg, Kinshasa, Lagos*. Ostfildern-Ruit, Hatje Cantz.

Ehrhardt, D. (2014). 'Indigeneity, belonging and religious freedom in Nigeria. Citizens' views from the street'. *Nigerian Research Network Policy Brief*, 5.

Ehrhardt, D. (2017). 'Indigeneship, bureaucratic discretion, and institutional change in Northern Nigeria'. *African Affairs*, 116 (464), 462–483.

Evans, I. (1997). *Bureaucracy and Race. Native Administration in South Africa*. Berkeley (CA), University of California Press.

Fabian, S. (2019). *Making Identity on the Swahili Coast: Urban Life, Community, and Belonging in Bagamoyo*. New York;, Cambridge University Press.

Fair, L. (2018). *Reel Pleasures. Cinema Audiences and Entrepreneurs in 20th Century Urban Tanzania*. Ohio University Press, Athens, Ohio.

Fall, B. and Roberts, R.L. (2019). 'Forced labor'. In: *General Labour History of Africa. Workers, Employers and Governments* (eds. S. Belluci and A. Eckert), 77–115.

James Currey, Suffolk, Boydell and Brewer, Rochester, International Labour Organisation, Abidjan.

Falola, T. (1985). 'From hospitality to hostility: Ibadan and strangers, 1830–1904'. *Journal of African History*, 26, 1.

Falola, T. (1989). *Politics and Economy in Ibadan, 1893–1945*. Lagos, Modelor.

Falola, T. (1995). 'Theft in colonial south-western Nigeria'. *Africa*, 50 (1), 10.

Falola, T. (2003). 'Slavery and pawnship in the Yoruba Economy of the XIXth Century'. In: *Pawnship, Slavery and Colonialism in Africa* (eds. P. Lovejoy and T. Falola). Trenton (NJ), Africa World Press.

Falola, T. and Oguntomisin, D. (1999). *Yoruba Warlords of the XIXth Century*. Trenton (NJ), Africa World Press.

Fassin, D. and Bourdelais, P. (dir.) (2005). *Les Constructions de l'intolérable. Études d'anthropologie et d'histoire sur les frontières de l'espace moral*. Paris, La Découverte.

Favarel-Garrigues, G. and Gayer, L. (2016). 'Violer la loi pour maintenir l'ordre. Le vigilantisme en débat'. *Politix*, 115 (3), 7–33.

Faye, O. and Thioub, I. (2003). 'Les marginaux et l'"État à Dakar'. *Le Mouvement Social*, 4, 93–108.

Feinstein, C.H. (2005). *An Economic History of South Africa, Conquest, Discrimination and Development*. Cambridge, Cambridge University Press.

Ferguson, J. (1999). *Expectations of Modernity: Myths and Meanings of Urban Life on the Zambian Copperbelt*. Berkeley (CA), University of California Press.

Ferguson, J. (2012). 'Theory from the Comaroffs, or how to know the world up, down, backwards and forwards'. *Cultural Anthropology*, www.culanth.org

Field, S. (2001). 'Windermere: Squatters, slumyars and removals, 1920–1960s'. In: *Lost Communities, Living Memories. Remembering Forced Removals in Cape Town* (ed. S. Field), 27–43. Le Cap, David Philip.

Fika, A.M. (1978). *The Kano Civil War and British Over-Rule, 1882–1940*. Oxford, Oxford University Press.

Fleisch, B.D. (1995). 'Social scientists as policy makers: E. G. Malherbe and the National Bureau for Educational and Social Research, 1929–1943'. *Journal of Southern African Studies*, 21 (3), 349–372.

Foucault, M. (1971). 'Nietzche, la généalogie, l'histoire'. In: *Dits et Ecrits I, 1954–1975*. Paris, Edition Gallimard.

Foucault, M. (1972). 'Revenir à l'histoire'. In: *Dits et Ecrits I, 1954–1975*. Paris, Edition Gallimard.

Foucault, M. (1977). 'Le jeu de Michel Foucault'. In: *Dits et Écrits II 1976–1988*. Paris, Gallimard.

Foucault, M. (1978). 'La gouvernementalité'. In: *Dits et Écrits II 1976–1988*. Paris, Gallimard.

Foucher, V. (2010). 'Achille Mbembe et l'hiver impérial français'. *Politique africaine*, 120 (4), 209–221.

Fourchard, L. (2001). *De la ville coloniale à la cour africaine. Espace, pouvoir et société à Ouagadougou et Bobo-Dioulasso*. Paris, L'Harmattan.

Fourchard, L. (2003). Urban Slums Report, a Case of Ibadan. Nigeria. www.ucl.ac.uk/dpu-projects/Global_Report/pdfs/Ibadan.pdf

Fourchard, L. (2005). 'Urban poverty, urban crime and crime control. The Lagos and Ibadan Cases, 1929–1945'. In: *African Urban Spaces in Historical Perspective* (eds. S. Salm and T. Falola), 291–319. Rochester (NY), University of Rochester Press.

Fourchard, L. (2006a). 'Lagos and the Invention of Juvenile Delinquency in Nigeria, 1920–1960'. *Journal of African History*, 47 (1), 115–137.

Fourchard, L. (2006b). 'Les territoires de la criminalité à Lagos et à Ibadan depuis les années 1930'. *Revue tiers monde*, 185.

Fourchard, L. (2007). 'Les rues de Lagos: Espaces disputés/espaces partagés'. *Flux*, 66–67, 64–65.

Fourchard, L. and Olukoju, A. (2007). 'State, local governments and the management of markets in Lagos and Ibadan since the 1950s'. In: *Gouverner les villes d'Afrique, État, gouvernement local et acteurs privés* (dir. Laurent Fourchard). Paris, Karthala.

Fourchard, L. (2008). 'A new name for an old practice, Vigilante in South Western Nigeria'. *Africa*, 78 (1), 16–40.

Fourchard, L. (2009). 'Dealing with 'strangers': Allocating urban space to migrants in Nigeria and French West Africa, End of the Nineteenth Century to 1960'. In: *African Cities. Competing Claims on Urban Spaces* (eds. F. Locatelli and P. Nugent), 187–218. Leyde, Brill.

Fourchard, L. (2010). 'The making of the juvenile delinquent in South Africa and Nigeria, 1930s–1970s'. *History Compass*, 7, 1–14.

Fourchard, L. (2011a). 'Between world history and state formation: New perspectives on Africa's cities'. *Journal of African History*, 52, 223–248.

Fourchard, L. (2011b). 'Lagos, Koolhaas and partisan politics in Nigeria'. *International Journal of Urban and Regional Research*, 35 (1), 40–56.

Fourchard, L. (2011c). 'The limits of penal reform: Punishing children and young offenders in South Africa and Nigeria (1930s–1960s)'. *Journal of Southern African Studies*, 37 (3), 517–534.

Fourchard L. (2011d). 'The politics of mobilisation for security in South African townships'. *African Affairs*, 110 (441), 607–627.

Fourchard, L. (2015). 'Bureaucrats and indigenes: Producing and bypassing certificates of origins in Nigeria'. *Africa*, 85 (1), 36–57.

Fourchard, L. (2016). 'Engagements sécuritaires et féminisation du vigilantisme en Afrique du Sud'. *Politix*, 115 (29), 57–78.

Fourchard, L. (2018). 'Citoyens d'origine contrôlée au Nigeria'. *Génèses*, 112 (3), 58–80.

Fourchard, L. and Albert, I.O. (dir.) (2003). *Sécurité, crime et ségrégation dans les villes d'Afrique de l'Ouest du XIXᵉ siècle à nos jours*. Ibadan, IFRA/Paris, Karthala.

Fourchard, L. and Segatti, A. (2015). 'Of xenophobia and citizenship: The politics of exclusion and inclusion in Africa'. *Africa*, 85 (1), 2–12.

Fouquet, T. (2014). Construire la blackness depuis l'Afrique, un renversement heuristique. *Politique africaine*, 136 (4), 5–19.

Forster-Towne, C. (2013). 'White hobby, black opportunity. Perceptions and motivations of police reservists in Johannesburg'. *South Africa Crime Quarterly*, 46.

Frankel, P. (2001). *An Ordinary Atrocity. Sharpeville and Its Massacre.* New Haven and London, Yale University Press.

French, K.J. (1983). *James Mpanza and the sofasonke Party in the Development of Local Politics in Soweto.* A dissertation submitted to the Faculty of Arts, Witwatersrand, Johannesburg.

Freund, B. (2007). *The African City: A History.* Cambridge, Cambridge University Press.

Gandy, M. (2006). 'Planning, anti-planning and the infrastructure crisis facing Metropolitan Lagos'. *Urban Studies,* 43 (2), 371–396.

Garland, D. (1997). 'Governmentality and the problem of crime: Foucault, criminology, sociology'. *Theoretical Criminology,* 1 (2), 173–214.

Garland, D. (2001). *The Culture of Control. Crime and Social Order in Contemporary Society.* Oxford, Oxford University Press.

Garland, D. (2014). 'What is a history of the present? On Foucault's genealogies and their critical preconditions'. *Punishment and Society,* 16 (4), 365–384.

Gary-Tounkara, D. (2008). *Migrants soudanais/maliens et conscience ivoirienne. Les étrangers en Côte d'Ivoire (1903–1980),* Paris, L'Harmattan.

Georges, A.A. (2011). 'Within salvation: Girl hawkers and the colonial state in development era Lagos'. *Journal of Social History,* 837–859.

Georges, A.A. (2014). *Making Modern Girls. A History of Childhood, Labor and Social Development in Colonial Lagos.* Athens, Ohio University Press.

Geschiere, P. (1995). *Sorcellerie et politique en Afrique. La viande des autres.* Paris, Karthala.

Geschiere, P. (2009). *The Perils of Belonging. Autochthony, Citizenship and Exclusion in Africa and Europe.* Chicago (IL), University of Chicago Press.

Geschiere, P. and Jackson, S. (2006). 'Autochthony and the crisis of citizenship: Democratization, decentralization, and the politics of belonging'. *African Studies Review,* 49 (2), 1–7.

Geschiere, P. and Nyamnjoh, F. (2000). 'Capitalism and autochthony: The seesaw and mobility of belonging'. *Public Culture,* 12 (2), 423–452.

Giliomee, H. (2003). *The Afrikaners, Biography of a People.* Le Cap, Tafelberg.

Ginisty, K. and Vivet, J. (2012). 'Territorialité d'un parti politique en ville. L'exemple du Frelimo à Maputo, capitale du Mozambique'. *L'espace politique,* 12 (3).

Glaser, C. (2000). *Bo-Tsotsi: The Youth Gangs of Soweto, 1935–1976.* Oxford, James Currey.

Glaser, C. (2005). 'Whistles and Sjamboks: Crime and Policing in Soweto, 1960–1976'. *South African Historical Journal,* 52 (1), 119–139.

Glaser, G.B. and Strauss, L.A. (1967). *The Discovery of Grounded Theory.* Chicago, Aldine.

Glasman, J. (2012). '"Connaître papier". Métiers de police et État colonial tardif au Togo'. *Genèses,* 86 (1), 37–54.

Glasman, J. (2014). 'Unruly agents: Police reform, bureaucratization and policemen's agency in interwar Togo'. *Journal of African History,* 55 (1), 79–100.

Glassman, J. (1995). *Feasts and Riots. Revelry, Rebellion, and Popular Consciousness on the Swahili Coast, 1856–1888*. Londres, James Currey.

Goerg, O. (1997). *Pouvoir colonial, municipalités et espaces urbains. Conakry- Freetown des années 1880 à 1914, t.2. Urbanisme et hygiénisme*. Paris, L'Harmattan.

Goffman, E. (1968). *Asiles. Études sur la condition sociale des malades mentaux*. Paris, Minuit.

Goldstein, D.M. (2005). 'Flexible justice: Neoliberal violence and "self-help" security in Bolivia'. *Critique of Anthropology*, 25 (4), 389–411.

Gondola, C.D. (1997). *Villes Miroirs. Migrations et identités urbaines à Kinshasa et Brazzaville*. Paris, L'Harmattan.

Goodhew, D. (1993). 'The people's police force: Communal policing initiatives in the western areas of Johannesburg, Circa 1930–1962'. *Journal of Southern African Studies*, 19 (3), 447–470.

Gooptu, N. (2001). *The Politics of the Urban Poor in Early Twentieth-Century India*, Cambridge, Cambridge University Press.

Graham, S. and Marvin, S. (2001). *Splintering Urbanism: Networked Infrastructures, Technological Mobilities and the Urban Condition*. Londres, Routledge.

de Gramont, D. (2015). 'Governing Lagos: Unlocking the politics of reforms'. Working paper, Carnegie Endowment for International Peace.

Gramsci, A. (1991). *Cahiers de Prison*. Paris, Gallimard.

Guichaoua, I. (2009). 'How do ethnic militias perpetuate in Nigeria? A micro-level perspective on the Oodua people's congress'. *World Development*, 38 (11), 1657–1666.

Guyer, J. (1995). 'Wealth in people, wealth in things'. *Journal of African History*, 36 (1), 83–90.

Guyer, J. and Salami, K. (2011). '"Kà s'ôwô": il n'y a pas d'argent!'. *Politique africaine*, 124 (4), 43–65.

Haenni, P. (2005). *L'ordre des caïds. Conjurer la dissidence urbaine au Caire*. Paris, Karthala/Le Caire, Cedej.

Haid, C.G. and Hilbrandt, H. (2019). 'Urban informality and the state: Geographical translations and conceptual alliances'. *International Journal of Urban and Regional Research*, 43 (3), 551–562.

Harnischfeger, J. 'The Bakassi boys: Fighting crime in Nigeria'. *Journal of Modern African Studies*, 41 (1), 2003, 23–49.

Hartog, F. (2003). *Régimes d'historicité. Présentisme et expériences du temps*. Paris, Seuil.

Haysom, N. (1986). 'Mabangalala: The rise of right-wing vigilantes in South Africa'. Working paper, Johannesburg, Wits University.

Heap, S. (2011). 'Their days are spent in gambling, loafing, pimping for prostitutes and picking pockets: Male juvenile delinquents on Lagos Island, 1920–1960s'. *Journal of Family History*, 35 (1), 48–70.

Hellman, E. (1940). *Problems of Urban Bantu Youth: Report of an Enquiry into the Causes of Early School-leaving and Occupational Opportunities Amongst Bantu*. Johannesburg, South African Institute of Race Relations.

Hellman, E. (1948). *Rooiyard: A Sociological Survey of an Urban Native Slum Yard*. Oxford, Oxford University Press.

Hendricks, C. and Musavengana, T. (2010). 'The security sector in Southern Africa'. Working paper, Pretoria, Institute for Security Studies.

Hibou, B. (1999a). 'De la privatisation des économies à la privatisation des États. Une analyse de la formation continue de l'État'. In: *La Privatisation des États* (dir. B. Hibou), 11–67. Paris, Karthala.

Hibou, B. (1999b). 'La "décharge", nouvel interventionnisme'. *Politique africaine*, 73 (1), 6–15.

Hibou, B. (2011). *Anatomie politique de la domination*. Paris, La Découverte.

Hibou, B. (2013). 'La bureaucratisation néolibérale ou la domination et le redéploiement de l'État dans le monde contemporain'. In: *La Bureaucratisation néolibérale* (dir. B. Hibou). Paris, La Découverte.

Hibou, B. (2017), Conference 'Travailler avec Weber', CERI, Sciences Po et Institut historique allemand, 4-6 octobre.

Higazi, A. (2007). 'Violence urbaine et politique à Jos (Nigeria), de la période coloniale aux élections de 2007'. *Politique africaine*, 106, 69–91.

Higazi, A. (2011). 'The Jos crisis. A recurrent Nigerian tragedy'. *Friedrich Ebert Stiftung. Discussion Paper* n° 2, January.

Higazi, A. (2016). 'Farmer-pastoralist conflicts on the Jos Plateau, central Nigeria: Security responses of local vigilantes and the Nigerian state'. *Conflict, Security, and Development*, 16 (4), 365–385.

Hilgers, M. (2012). 'The historicity of the neoliberal State'. *Social Anthropology*, 20 (1), 80–94.

Hills, A. (2008). 'The dialectic of police reform in Nigeria'. *Journal of Modern African Studies*, 46 (2), 215–234.

Hindson, D. (1987). *Pass Controls and the Urban African Proletariat*. Johannesburg, Ravan Press.

Hoffmann, K. and Kirk, T. (2013). Public authority and the provision of public goods in conflict-affected and transitioning regions. *The Justice and Security Research Program, Paper* n° 7.

Home, R. (1997). *Of Planting and Planning. The Making of British Colonial Cities*. London, E. et F. N. Spon/Chapman Hill.

Hull, M.S. (2012a). *Government of Paper: The Materiality of Bureaucracy in Urban Pakistan*. Berkeley, University of California Press.

Hull, M.S. (2012b). 'Documents and bureaucracy'. *Annual Review of Anthropology*, 41, 251–267.

Human Rights Watch (2006). '"They do not own this place". Government Discrimination Against "Non-Indigenes" in Nigeria. Working paper, Human Rights Watch, www.hrw.org.

Idowu, S. (1980). *Armed Robbery in Nigeria*. Lagos, Jacob and Johnson Books.

Ige, B. (1995). *People, Politics and Politicians of Nigeria (1940–1979)*. Ibadan, Heinemann.

Ikelegbe, A. (2001). 'The perverse manifestation of civil society: Evidence from Nigeria'. *Journal of Modern African Studies*, 39 (1), 1–24.

Illife, J. (1987). *The African Poor: A History*, Cambridge, Cambridge University Press.

International Crisis Group, ICG (2017). Watchmen of Lake Chad: Vigilante Groups Fighting Boko Haram. Report n° 244, 23 February.

Inyang, E.O. (1989). 'The Nigeria police force: Peace keeping in Nigeria'. In: *Nigeria since Independence. The First 25 Years*, vol. 4 (eds. T.N. Tamuno and J.A. Atanda). Ibadan, Heinemann.

Ismail, O. (2009). 'The dialectic of "junctions" and "bases": Youth "securo-commerce" and the crisis of order in downtown Lagos'. *Security Dialogue*, 4 (4–5), 463–487.

Jackson, S. (2006). 'Sons of which soils? The language and politics of autochthony in Eastern D. R. Congo'. *African Studies Review*, 49 (2), 95–123.

Jaffe, R. (2012). 'Criminal dons and extralegal security privatization in downtown Kingston, Jamaica'. *Singapore Journal of Tropical Geography*, 33 (2), 184–197.

Jedlowski, A. (2012). 'Small screen cinema: Informality and remediation in Nolly-wood'. *Television & New Media*, 13 (5), 431–446.

Jemibewon, D.M. (2001). *The Nigeria Police in Transition: Issues, Problems and Prospects*. Spectrum, Ibadan.

Jennings, M. and Mercer, C. (2011). 'Réhabiliter les nationalismes: Convivialité et conscience nationale en Tanzanie post-coloniale'. *Politique africaine*, 121, 87–106.

Jensen, S. (2005). 'Above the law: Practices of sovereignty in Surrey Estate, Cape Town'. In: *Sovereign Bodies. Citizens, Migrants and States in the Postcolonial World* (eds. T. Blom Hansen and F. Stepputat), 218–238. Princeton (NJ), Princeton University Press.

Jensen, S. (2008). *Gangs, Politics and Dignity in Cape Town*. Oxford, James Currey.

Jobard, F. and De Mailard, J. (2015). *Sociologie de la police. Politiques, organisations, réformes*. Paris, Armand Colin.

Johnston, L. (1996). 'What is vigilantism?'. *British Journal of Criminology*, 36 (2), 220–236.

Joseph, R. (1987). *Democracy and Prebendal Politics in Nigeria: The Rise and Fall of the Second Republic*. Cambridge, Cambridge University Press.

Joseph, R. (2013). 'Industrial policies and contemporary Africa: The transition from prebendal to developmental governance'. In: *The Industrial Policy Revolution. Africa in the Twenty-first Century*, t. 2 (eds. J. Stiglitz, J. Yifu Lin, and J. Esteban). Basingstoke, Palgrave Macmillan.

Junod, H.P. (1948). 'The prevention of crime and the right treatment of delinquents'. *Penal Reform News*, 5.

Kafka, B. (2009). 'Paperwork: The state of the discipline'. *Book History*, 12, 340–353.

Kaldenberg, J. (1951). Acquisition and Extinction of the Anti-Social Habits of Two Hundred Pupils at Constantia Reformatory. Unpublished PhD thesis, University of Cape Town.

Kapp, C. (2008). 'Crystal meth boom adds to South Africa's health challenges'. *The Lancet*, 371, 9608, January, 193–194.

Keese, A. (2014). 'Slow abolition within the colonial mind: British and French debates about "vagrancy", "African laziness", and forced labour in West Central and South-Central Africa, 1945–1965'. *International Review of Social History*, 59 (3), 377–407.

Kidambi, P. (2004). '"The ultimate masters of the city": Police, public order and the poor in colonial Bombay, c. 1893–1914'. *Crime, histoire et sociétés*, 8 (1), 2–47.

King, D. and Le Galès, P. (2011). 'Sociologie de l'État en recomposition'. *Revue française de sociologie*, 52 (3), 453–480.

Kinnes, I. (2000). 'From urban street gangs to criminal empires: The changing face of gangs in the Western Cape'. Working paper, Pretoria, Institute for Security Studies, p. 22.

Kirk-Greene. A. (1968). *Lugard and the Amalgamation of Nigeria: A Documentary Record*. Londres, Franck Cass and Co.

Kirk-Greene. A. (1983). 'Ethnic engineering and the federal character of Nigeria: Boon of contentment or bone of contention'. *Ethnic and Racial Studies*, 6 (4), 457–476.

Kirsch, T.G. (2010). 'Violence in the name of democracy. Community policing, vigilante action and nation-building in South Africa'. In: *Domesticating Vigilantism in Africa* (eds. T.G. Kirsch and T. Gratz). Oxford, James Currey.

Kirsch, T.G. and Grâtz, T. (eds.) (2010). *Domesticating Vigilantism in Africa*. Oxford, James Currey.

Koolhaas, R., et al. (2000). 'Lagos, Harvard project on the city'. In: Arc en Rêve centre d'architecture (dir.), *Mutations, événement culturel sur la ville contemporaine*. Bordeaux, Arc en rêve.

Koolhaas, R. (2002). 'Fragments of a lecture on Lagos'. In: *Under Siege: Four African Cities: Freetown, Johannesbourg, Kinshasa, Lagos* (eds. Okwui Enwezor et al.). Ostfildern-Ruit, Hatje Cantz.

Koonings, K. and Kruijt, D. (eds.) (2009). *Megacities: The Politics of Urban Exclusion and Violence in the Global South*. London, Zed Books.

Krause, J. (2011). A Deadly Cycle: Ethno-Religious Conflict in Jos, Plateau State, Nigeria. Working Paper, Geneva Declaration Secretariat, Geneva.

Krause, J. (2017). 'Non-violence and civilian agency in communal war: Evidence from Jos, Nigeria'. *African Affairs* 116, 463, 261–283.

Kuba, R. and Lentz, C. (eds.) (2006). *Land and the Politics of Belonging in West Africa*. Leyde, Brill.

Kynoch, G. (2003). 'Friend or foe? A world view of community-police relations in Gauteng Townships, 1947–1977'. *Canadian Journal of African Studies*, 37 (2–3), 298–327.

Kynoch, G. (2005a). *We are Fighting the World. A History of the Marashea Gangs in South Africa, 1947–1999*. Athens (Ohio), Ohio University Press/Pietermaritzburg, University of KwaZulu Natal Press.

Kynoch, G. (2005b). 'Crime, conflict and politics in transition-era South Africa'. *African Affairs*, 104 (416), 493–514.

Kynoch, G. (2008). 'Urban violence in colonial Africa: A case for South African exceptionalism'. *Journal of Southern African Studies*, 34 (3), 629–645.

Kynoch, G. (2011). 'Of compounds and cellblocks: The foundations of violence in Johannesburg, 1890–1950'. *Journal of Southern African Studies*, 37 (3), 463–477.

La Hausse, P. (1988). *Brewers, Beerhalls and Boycotts: A Study of Liquor in South Africa*. Johannesburg, Ravan Press.

Landau, L.B. (ed.) (2011). *Exorcising the Demons Within: Xenophobia, Violence, and Statecraft in Contemporary South Africa*. Johannesburg, Wits University Press.

Lar, J. (2015). Vigilantism, State, and Society in Plateau State, Nigeria. A History of Plural Policing (1950 to the Present). Phd, University of Bayreuth.

Lar, J. (2017). 'Historicising vigilante policing in Plateau State, Nigeria'. In: *Police in Africa: The Street Level View*' (eds. J. Beek, M. Gopfert, O. Owen, and J. Steinberg). Londres, Hurst.

Lascoumes, P. and Le Galès, P. 'L'action publique saisie par ses instruments'. In: *Gouverner par les instruments* (dir. P. Lascoumes and P. Le Galès), 11–44. Paris, Presses de Sciences Po, 2004.

Last, M. (2000). 'Children and the experience of violence: Contrasting cultures of punishment in Northern Nigeria'. *Africa*, 70 (3), 359–393.

Le Galès, P. (2003). *Le Retour des villes européennes. Sociétés urbaines, mondialisation, gouvernement et gouvernance*. Paris, Presses de Science Po.

Le Galès, P. (2016). 'Neoliberalism and urban change: Stretching a good idea too far?'. *Territory, Politics, Governance*, 4 (2), 154–172.

Le Galès, P. (2019). 'Pourquoi si peu de comparaison en sociologie urbaine?' In: *D'une ville à l'autre. La comparaison internationale en sociologie urbaine* (eds. J.Y. Autier, V. Baggioni, B. Cousin, Y. Faijalkow and L. Launay), 21–41. Paris, La Découverte.

Le Galès, P. (2020). 'A la recherche du politique dans les villes'. *Raisons politiques*, 79 (3), 11–40.

Lee, R. (2009). *African Women and Apartheid: Migration and Settlement in Urban South Africa*. Londres, I. B. Tauris.

Legassick, M. (2002). Armed Struggle and Democracy. The Case of South Africa. Discussion paper 20. Nordiska Afrikainstitutet, Uppsala.

Lemanski, C. (2004). 'A new apartheid? The spatial implications of fear of crime in Cape Town, South Africa'. *Environment and Urbanization*, 16 (2), 101–112.

Lemon, A. (ed.) (1991). *Homes Apart. South Africa's Segregated Cities*. Bloomington (IN), Indiana University Press.

Lentz, C. (2006). 'Land rights and the politics of belonging in Africa: An introduction'. In: *Land and the Politics of Belonging in West Africa* (eds. K. Richard and C. Lentz), 1–34. Leyde, Brill.

Lentz, C. and Nugent, P. (eds.) (2000). *Ethnicity in Ghana. The Limits of Invention*. Basingstoke-Houndmills-London, Macmillan Press, New York;, St. Martin's Press.

Lewis, J. (2000). *Empire State-Building. War and Welfare in Kenya, 1925–1952*. Oxford, James Currey/Nairobi, EAEP/Athens (Ohio), Ohio University Press.

Limoncelli, S. (2010). *The Politics of Trafficking: The First International Movement to Combat the Sexual Exploitation of Women*. Redwood City (CA), Stanford University Press.

Lindell, I. (2008). 'The multiple sites of urban governance: Insights from an African city'. *Urban Studies*, 45—9, 1879—1901.

Lindell, I. and Utas, M. (2012). 'Networked city life in Africa'. *Urban Forum*, 23, 409–414.

Lindsay, L.A. (2003). *Working with Gender: Wage Labor and Social Change in South-Western Nigeria*. Portsmouth (NH), Heinemann.

Lipsky, M. (1980). *Street Level Bureaucracy: Dilemmas of the Individual in Public Services*. Russell Sage Foundation.

Locatelli, F. and Nugent, P. (2009). *African Cities. Completing Claims on Urban Spaces*. Leiden and London Brill.

Lodge, T. (2011). *Sharpeville. An Apartheid Massacre and its Consequences*. Oxford, Oxford University Press.

Lonsdale, J. (1981). 'States and social processes in Africa: A historiographical survey'. *African Studies Review*, 24 (2–3), 139–225.

Lovejoy, P. (1980). *The Caravans of Kola: The Hausa Kola Trade, 1700–1900*. Evanston (IL), Northwestern University Press.

Lubeck, P. (1985). 'Islamic protest under semi-industrial capitalism: Yan Tatsine explained'. *Africa*, 55 (4), 369–389.

Lüdtke, A. (2000). *Des ouvriers dans l'Allemagne du XXᵉ siècle. Le quotidien des dictatures*. Paris, L'Harmattan.

Lüdtke, A. (2015a). 'La domination comme pratique sociale'. *Sociétés contemporaines*, 99–100 (3), 49.

Lüdtke, A. (2015b). 'L'Histoire comme science sociale. Entretien de Oeser Alexandra avec Alf Lüdtke'. *Sociétés contemporaines*, 99–100 (3), 169–191.

Lukpata, V. I. (2013). 'Revenue allocation formulae in Nigeria: A continuous search'. *International Journal of Public Administration and Management Research*, 2 (1), October, 32–38.

Lund, C. (2006). 'Twilight institutions: Public authority and local politics in Africa'. *Development and Change*, 37 (4), 685–705.

Lynn, M. (2006). 'The Nigerian self-government crisis of 1953 and the Colonial Office'. *The Journal of Imperial and Commonwealth History*, 34 (2), 245–261.

Mabogunje, A.L. (1962). 'The growth of residential districts in Ibadan'. *Geographical Review*, 1, 56–77.

Mabogunje, A.L. (1968). *Urbanisation in Nigeria*. London, University of London Press.

Madueke, K.L. (2017). 'From neighbors to deadly enemies: Excavating landscapes of territoriality and ethnic violence in Jos, Nigeria'. *Journal of Contemporary African Studies*, 36 (1), 87–102.

Madueke, K.L. (2018). 'Routing ethnic violence in a divided city: Walking in the footsteps of armed mobs in Jos, Nigeria'. *Journal of Modern African Studies*, 56 (3), 443–470.

Madueke, K.L. (2019). 'The emergence and development of ethnic strongholds and frontiers of collective violence in Jos, Nigeria'. *African Studies Review*, 62 (4), 6–30.

Madueke, K.L. and Vermeulen, F.F. (2018). 'Frontiers of ethnic brutality in an African city: Explaining the spread and recurrence of violent conflict in Jos, Nigeria'. *Africa Spectrum*, 53 (2), 37–63.

Malaquais, D. (dir.) (2006). 'Cosmopolis: De la ville, de l'Afrique et du monde'. *Politique africaine*, 10017–37.

Malherbe, E.G. (1932). 'Education and the Poor White'. vol. 3 *Report of the Carnegie Commission: The Poor White Problem in South Africa*. Stellenbosch, Pro Eclesia Drukkery.

Malherbe, E.G. (1977). *Education in South Africa, 1925–1977*, 316–317. Le Cap, Juta and Co.

Mamdani, M. (1996). *Citizen and Subject: Contemporary Africa and the Legacy of Late Colonialism*. Princeton (NJ), Princeton University Press.

Mann, G. (2009). 'What was the Indigénat? The "Empire of Law" in French West Africa'. *Journal of African History*, 50 (3), 331–353.

March, W.L. (1946). '*The problem of the child delinquent, Johannesburg*'. *Penal Reform League of South Africa*, Pretoria.

Marenin, O. (1987). 'The Anini Saga: Armed robbery and the reproduction of ideology in Nigeria'. *The Journal of Modern African Studies*, 25 (2), 259–281.

Marks, M., Shearing, C. and Wood, J. (2009). 'What should the police be? Finding a new narrative for community policing in South Africa'. *Police Practice and Research*, 10 (2), 145–155.

Marshall-Fratani, R. (2006). 'The war of "who is who". Autochthony, nationalism and citizenship in the Ivoirian crisis'. *African Studies Review*, 49 (2), 9–43.

Martin, D-C. (1998). 'Le poids du nom. Culture populaire et constructions identitaires chez les "métis" du Cap'. *Critique Internationale*, 1, 73–100.

Martin, P.M. (2002). *Leisure and Society in Colonial Brazzaville*. Cambridge, Cambridge University Press.

Mattina, C. (2016). *Clientélismes urbains. Gouvernement et hégémonie politique à Marseille*. Paris, Presses de Sciences Po.

Mayer, P. (1961). *Townsmen and Tribesmen; Conservatism and the Process of Urbanization in a South African City*. Oxford, Oxford University Press.

Maylam, P. (1995). 'Explaining the apartheid city: 20 years of South African urban historiography', *Journal of Southern African Studies*, 21 (1), 19–38.

Mbembe, A. (1999). 'Du gouvernement privé indirect'. *Politique africaine*, 73, 103–122.

Mbembe, A. (2005) [2000]. *De la postcolonie. Essai sur l'imagination politique dans l'Afrique contemporaine*. Paris, Karthala.

Mbembe, A. (2010). 'Faut-il provincialiser la France?'. *Politique africaine*, 119 (3), 155–188.

McDonnell, E.M. (2017). 'Patchwork Leviathan: How pockets of bureaucratic governance flourish within institutionally diverse developing states'. *American Sociological Review*, 82 (3), 476–510.

McFarlane, C. (2008). 'Governing the contaminated city: Infrastructure and sanitation in colonial and post-colonial Bombay'. *International Journal of Urban and Regional Research*, 32 (2), 415–435.

McFarlane, C. (2010). 'The comparative city: Knowledge, learning, urbanism'. *International Journal of Urban and Regional Research*, 34 (4), 725–742.

McLachlan, F. (1986). *Children, their Courts and Institutions in South Africa*. Institute of Criminology, University of Cape Town.

Meagher, K. (2007). 'Hijacking civil society: The inside story of the Bakassi Boys vigilante group of south-eastern Nigeria'. *Journal of Modern African Studies*, 45 (1), 89–115.

Meagher, K. (2018). 'Taxing times: Taxation, divided societies and the informal economy in Northern Nigeria'. *The Journal of Development Studies*, 54 (1), 1–17.

Médard, J.-F. (1991). *États d'Afrique noire. Formation, mécanisme et crise*. Paris, Karthala.

Midgley, J. (1982). 'Corporal punishment and penal policy: Notes on the continued use of corporal punishment with reference to South Africa'. *Journal of Criminal Law and Criminology*, 73 (1), 395.

Milner, A. (1972). *The Nigerian Penal System*. London, Sweet and Maxwell.

Miraftab, F. (2004). 'Making neo-liberal governance: The disempowering work of empowerment'. *International Planning Studies*, 9 (4), 239–259.

Mitchell, T. (1991). 'The limits of the state: Beyond statist approaches and their critics'. *American Political Science Review*, 85 (1), 77–99.

Moore, M. Prichard Wilson and Fjeldstard Odd-Helge. (2018). *Taxing Africa: Coercion, Reform and Development*. London, Zed Books.

Monson, T. (2015). 'Everyday politics and collective mobilization against foreigners in a South African shack settlement'. *Africa*, 85 (1), 131–153.

Montagus, A. (1975). *Race and IQ*. Oxford, Oxford University Press.

Moodie, D. (2005). 'Maximum average violence: Underground assaults on the South African goldmines, 1900–1950'. *Journal of Southern African Studies*, 31 (3), 547–567.

Morange, M. (2015). Street trade, neoliberalisation and the control of space: Nairobi's Central Business District in the era of entrepreneurial urbanism. *Journal of Eastern African Studies*, 9 (2), 247–269.

Morelle, M. (2007). *La Rue des enfants. Les enfants des rues Yaoundé et Antanarivo*. Paris, CNRS Éditions.

Morton, D. (2019). *Age of Concrete: Housing and the Shape of Aspiration in the Capital of Mozambique*. Ohio University Press.

Muchielli, L. (dir.). (2008). *La Frénésie sécuritaire. Retour à l'ordre et nouveau contrôle social*. Paris, La Découverte.

Müller, M.M. (2016). *The Punitive City, Privatized Policing and Protection in Neoliberal Mexico*. Zed Books, London.

Muncie, J. (1999). *Youth and Crime: A Critical Introduction*. Londres, Sage.

Musemwa, M. (1996). 'Administering an African township or a personal fiefdom? The management style of the manager of Langa township, Cape Town, 1948–1948'. *Kleio*, 28, 1, 137–152.

Mustapha, A. R. (2007). 'Institutionalizing ethnic representation: How effective is the Federal Commission in Nigeria'. CRISe working paper, 43.

Mustapha, A.R. and Ehrhardt, D. (eds.) (2018). *Creed and Grievance: Muslim-Christian Relations and Conflict Resolution in northern Nigeria*. Oxford, James Currey, Boydell and Brewer Ltd.

Mustapha, A.R., Higazi, A., Lar J. and Chromy, K. (2018). 'Jos: A decade of fear and violence in central Nigeria'. In: *Creed and Grievance: Muslim-Christian Relations and Conflict Resolution in northern Nigeria* (eds. A.R. Mustapha and D. Ehrhardt). Oxford, James Currey, Boydell and Brewer Ltd.

Muthien, Y. (1994). *State and Resistance in South Africa, 1939–1965*. Aldershot, Avebury Publisher.

Mutongi, K. (2006). 'Thugs or entrepreneurs? Perceptions of *Matatu* operators in Nairobi, 1970 to the present'. *Africa*, 76 (4), 549–568.

Myers, G.A. (2003). *Verandas of Power. Colonialism and Space in Urban Africa*. Syracuse (NY), Syracuse University Press.

Myers, G. (2011). *African Cities. Alternative Visions of Urban Theory and Practice*. Londres/New York; (NY), Zed Books.

Naanen, B.B. (1991). 'Itinerant gold mines: Prostitution in the cross river basin of Nigeria, 1930–1950'. *African Studies Review*, 34, 57–79.

Neocosmos, M. (2010). *From 'Foreign Natives' to 'Native Foreigners': Explaining Xenophobia in Post-Apartheid South Africa: Citizenship and Nationalism, Identity and Politics*. Dakar, Senegal: CODESRIA.

Ngalamulume, K. (2004). 'Keeping the city totally clean: Yellow fever and the politics of prevention in colonial Saint-Louis-du-Sénégal, 1850–1914'. *Journal of African History*, 54, 183–202.

Nieftagodien, N. (2011). 'Xenophobia's local genesis: Historical constructions of insiders and the politics of exclusion in Alexandra Township'. In: L.B. Landau (ed.), *Exorcising the Demons Within: Xenophobia, Violence, and Statecraft in Contemporary South Africa*, 109–134. Johannesburg, Wits University Press.

Nigeria's Governor Forum (2016). 'Internally Generated Revenue of Nigerian States – Trends, Challenges and Options'. *Working paper*.

Nightingale, C. H. (2012). *Segregation: A Global History of Divided City*. Chicago (IL), University of Chicago Press.

Nolte, I. (2007). 'Ethnic vigilantes and the State: The Oodua people's congress in South-Western Nigeria'. *International Relations*, 21, 217–235.

Nolte, I. (2008). '"Without women, nothing can succeed": Yoruba women in the Oodua People's Congress (OPC), Nigeria'. *Africa*, 78 (1), 84–106.

Nolte, I. (2009). *Obafemi Awolowo and the Making of Remo: The Local Politics of a Nigerian Nationalist*. Édimbourg, Edinburgh University Press.

Nolte, I. and Hoffmann, L. (2013). 'The roots of neopatrimonialism: Opposition politics and popular consent in southwest Nigeria'. In: *Democracy and Prebendalism in Nigeria. Critical Interpretations* (eds. W. Adebanwi and E. Obadare). Basingstoke, Palgrave Macmillan.

Northern Region of Nigeria (1953). *Report on the Kano Disturbances, 16–19 May, 1953*. Kaduna, Government Printer.

Nuttall, S. and Mbembe, A. (2004). 'Writing the world from an African metropolis', *Public Culture*, 16 (3), 347–372.

Nuttall, S. and Mbembe, A. (eds.) (2009). *Johannesburg: The Elusive Metropolis*. Johannesburg, Wits University Press/Durham (NC), Duke University Press.

Nuttall, S. and Mbembe, A. (2005). 'Blasé attitude: A response to Michael Watts'. *Public Culture*, 17 (1), 193–202.

Obadare, E. (2007). 'Lamidi Adedibu ou l'État nigérian entre contraction et sous-traitance'. *Politique africaine*, 106 (2), 117–118.

Obbi, E. (1989). 'Crime and delinquency in metropolitan lagos: A study of 'crime and delinquency area' theory'. *Social Forces*, 67 (3), 751–765.

Okpara, E.E. (1988). 'The role of touts in passenger transport in Nigeria'. *The Journal of Modern African Studies*, 26 (2), 327–335.

Olaniyi, R. (2003). 'Nationalist movement in a multiethnic community of Sabon-Gari, Kano'. In: *Perspectives on Kano-British Relations* (ed. M.O. Hambolu). Kano, Gidan Museum.

Olaniyi, R. (2005a). 'Yoruba commercial diaspora and settlement patterns in Pre-Colonial Kano'. In: *Nigerian Cities* (eds. S. Salm and T. Falola), 89–92. Trenton (NJ), Africa World Press.

Olaniyi, R. (2005b). *Community Vigilantes in Metropolitan Kano, 1985–2005*. Ibadan, IFRA.

Olaniyi, R. (2013). 'Approaching the study of Yoruba diaspora in Northern Nigeria in the xxth century'. *IFRA Special research Issue*, 2, 67–89

Olukoju, A. (1991). 'Prohibition and paternalism: The State and the clandestine liquor traffic in Northern Nigeria, c. 1898–1918'. *The International Journal of African Historical Studies*, 24 (2), 349–368.

Olukoju, A. (2003). 'The segregation of Europeans and Africans in Colonial Nigeria'. In: *Security, Crime and Segregation in West African Cities since the Nineteenth Century* (eds. L. Fourchard and I.O. Albert). Paris, Karthala/Ibadan, IFRA.

Omobowale, A.O. and Olutayo, A. (2007). 'Chief Lamidi Adedibu and patronage politics in Nigeria'. *Journal of Modern African Studies*, 45 (3), 425–446.

Omobowale, A.O. and Fayiga, O.O (2017). 'Commercial motor drivers, transport unions and electoral violence in Ibadan, Nigeria'. *Development and Society*, 46 (3), 591–614.

Oneyonoru, I. (2003). 'Insecurity and the 'Bakassi Boys' operations in Eastern Nigeria'. In: *Sécurité, crime et ségrégation dans les villes d'Afrique de l'Ouest du XIXᵉ siècle à nos jours* (dir. L. Fourchard and I.O. Albert), 380–381. Paris, Karthala/Ibadan, IFRA.

Osaghae, E. (1991). 'Ethnic minorities and federalism in Nigeria'. *African Affairs*, 90 (359), 237–258.

Osaghae, E. (1993). *Trends in Migrant Political Organisations in Nigeria, the Igbo in Kano*. Ibadan, IFRA.

Osaghae, E. (1998). *Crippled Giant: Nigeria since Independence*. Londres, Hurst and Co.

Ostien, P. (2009). Jonah Jang and the Jasawa: Ethno-religious conflict in Jos, Nigeria. *Muslim Christian Relations in Africa*, www.sharia-in-africa.net.

Owen, O. (2012). 'Maintenir l'ordre au Nigeria: vers une histoire de la souveraineté de l'État'. *Politique africaine*, 128 (4), 25–51.

Owen, O. and Cooper-Knock, S.-J. (2015). 'Between vigilantism and bureaucracy: Improving our understanding of police work in Nigeria and South Africa'. *Theoretical Criminology*, 19 (3), 355–375.

Owen, O. and Goodfellow, T. (2018). 'Taxation, property rights and the social contract in Lagos'. *ICTD Working Paper 73*.

Oyesiku, K. (1998). *Modern Urban and Regional Planning: Law and Administration in Nigeria*. Ibadan, Kraft Books.

Packer, G. (2006). 'The Megacity. Decoding the chaos of Lagos'. *The New Yorker*, November 13.

Paden, J. (1986). *Ahmadu Bello, Sardauna of Sokoto: Values and Leadership in Nigeria*. Portsmouth, Hodder and Stoughton.

Paden, J. (1973). *Religion and Political Culture in Kano*. Berkeley (CA), University of California Press.

Palonen, K. (2003). 'Four times of politics: Policy, polity, politicking, and politicization'. *Alternatives*, 29, 171–186.

Parnell, S. (2003). 'Race, power and urban control: Johannesburg's Inner City Slum-Yards, 1910–1923'. *Journal of Southern African Studies*, 29, 615–637.

Parnell, S. and Mabin, A. 'Rethinking urban South Africa'. *Journal of Southern African Studies*, 2 (1), 1995, 39–61.

Paton, A. (1986). *Diepkloof. Reflections of Diepkloof Reformatory*. Le Cap, David Philip.

Paton, A. (1948). *Freedom as a Reformatory Instrument*. Pretoria, Penal Reform League of South Africa.

Péclard, D. and Hagmann, T. (2010). 'Negotiating statehood: Dynamics of power and domination in Africa'. *Development and Change*, 41 (4), 539–562.

Peck, J. and Tickell, A. (2002). 'Neoliberalizing space'. *Antipode*, 34 (3), 380–404.

Peel, J.D. (1978). Olaju, A Yoruba concept of development'. *The Journal of Development Studies*, 12 (2)p. 139–165.

Peel, J.D. (2011). 'Un siècle d'interaction entre islam et christianisme dans l'espace yoruba'. *Politique africaine*, 123, 27–50.

Pinnock, D. (1984). *The Brotherhood: Street Gangs and State Control in Cape Town*. Le Cap, Philip David.

Pinson, G. and Morel Journel, C. (2016). 'The neoliberal city – Theory, evidence, debates'. *Territory, Politics, Governance*, 4 (2), 137–153.

Pieterse, E. and Parnell, S. (2016). 'Translational global praxis: Rethinking methods and modes of African urban research'. *International Journal of Urban and Regional Research*, 40 (1), 236–246.

Pitcher, A., Moran M.H. and Johnson, M. (2009). 'Rethinking patrimonialism and neopatrimonialism in Africa'. *African Studies Review*, 52 (1), 126–156.

Plotnicov, L. (1967). *Strangers to the City. Urban Man in Jos, Nigeria*. Pittsburgh (PA), University of Pittsburgh Press.

Pogrund, B. (1990). *Sobukwe and Apartheid*. New York; (NY), Peter Halban.

Posel, D. (1991). *The Making of Apartheid, 1948–1961, Conflict and Compromise*. Oxford, Oxford University Press.

Posel, D. (2005). 'The case for a Welfare State: Poverty and the politics of the urban African family in the 1930s and 1940s'. In: *South Africa's 1940s: World of Possibilities* (eds. S. Dubow and A. Jeeves). Le Cap, Double Storey Books.

Potts, D. (2006). '"Restoring order"? Operation Murambatsvina and the urban crisis in Zimbabwe'. *Journal of Southern African Studies*, 32 (2), 273–291.

Potts, D. (2012). 'Challenging the myths of urban dynamics in Sub-Saharan Africa: The evidence from Nigeria'. *World Development*, 40 (7), 1382–1393.

Pratten, D. (2007). 'Singing thieves: History and practice in Nigerian popular justice'. In: *Global Vigilantes* (eds. D. Pratten and A. Sen), 175–206. Londres, Hurst and Co.

Pratten, D. and Sen, A. (2007). 'Global vigilantes: Perspectives on justice and violence'. In: *Global Vigilantes* (eds. D. Pratten and A. Sen). Londres, Hurst and Co.

Ramphele, M. (1993). *A Bed Called Home: Life in the Migrant Labour Hostels of Cape Town*. Le Cap, David Philip/Athens (Ohio), Ohio University Press/ Édimbourg, Edinburgh University Press.

Rancière, J. (1987). *Le Maître ignorant. Cinq leçons sur l'émancipation intellectuelle*. Paris, Fayard.

Rancière, J. (2005). 'L'actualité du Maitre ignorant' entretien avec Jacques Rancière réalisé par Andréa Benvenuto, Laurence Cornu et Patrice Vermeren. *Le Télémaque*, 2 (27), 21–36.

Ranger, T. (2010). *Bulawayo Burning. The Social History of a Southern African City, 1893-1960*. Oxford, James Currey.

Rasmussen, J. (2012). 'Inside the system, outside the law: Operating the Matatu sector in Nairobi'. *Urban Forum*, 23, 415–432.

Reid, R. (2011). 'Past and presentism, the 'precolonial' and the foreshortening of African history'. *Journal of African History*, 52 (2), 135–155.

Revilla, L. (2020). Volontaires de l'ordre, police du quotidien dans les quartiers populaires de Lagos et Khartoum. PhD in Political Science, Sciences Po Bordeaux.

Rich, P. (1986). *White Power and the Liberal Conscience: Racial Segregation and South African Liberation, 1921–1960*. Manchester, Manchester University Press.

Rizzo, M. (2011). 'Life is war. Informal transport workers and neoliberalism in Tanzania, 1998-2009'. *Development and Change*, 42 (4), 1179–1206.

Robert, P. (2005). *La Sociologie du crime*. Paris, La Découverte.

Robinson, J. (1991). 'Administrative strategies and political power in South Africa's black townships, 1930–1960'. *Urban Forum*, 2 (2), 63–77.

Robinson, J. (1996). *The Power of Apartheid: State, Power and Space in South African Cities*. Butterworth-Heinemann.

Robinson, J. (2006). *Ordinary Cities. Between Modernity and Development*. Londres.

Robinson, J. (2011). 'Cities in a world of cities: The comparative gesture'. *International Journal of Urban and Regional Research*, 35 (1), 1–23.

Robinson, J. (2012). '(Re)theorizing cities from the South. Looking beyond Neo-liberalism'. *Urban Geography*, 33 (4), 593–617.

Robinson, J. (2016). 'Thinking cities through elsewhere: Comparative tactics for a more global urban studies'. *Progress in Human Geography*, 40 (1), 3–29.

Robinson, J. and Roy, A. (2016). 'Debate on global urbanisms and the nature of urban theory'. *International Journal of Urban and Regional*, 40 (1), 181–186.

Rodgers, D. (2019). 'From 'broder' to 'don': Methodological reflections on longitudinal gang research in Nicaragua'. In: *Ethnography as Risky Business: Field Research in Violent and Sensitive Contexts* (eds. K. Koonings, D. Kruijt and D. Rodgers). Lanham, MA: Lexington Press.

Roelofs, P. (2016). The Lagos Model and the Politics of Competing Conceptions of Good Governance in Oyo State, Nigeria, 2011–2015. PhD, London School of Economics.

Roelofs, P. (2019). 'Beyond programmatic versus patrimonial politics: Contested conceptions of legitimate distribution in Nigeria'. *The Journal of Modern African Studies*, 57 (3), 415–436.

Rosenbaum, H. J. and Sederberg, P. C. (1974). 'Vigilantism: An analysis of establishment violence'. *Comparative Politics*, 6 (4), 541–570.

Rotimi, K. (2001). *The Police in a Federal State. The Nigerian Experience*. Ibadan, College Press.

Roy, A. (2009). 'Why India cannot plan its cities: Informality, insurgence and the idiom of 2009 urbanization'. *Planning Theory*, 8 (1), 76–87.

Roy, A. (2011). 'Slumdog cities, Rethinking subaltern urbanism'. *International Journal of Urban and Regional Research*, 35 (2), 223–232.

Roy, A. (2016). 'Who is afraid of postcolonial theory'. *International Journal of Urban and Regional Research*, 40 (1), 200–209.

Rubbers, B. and Roy, A. (2015). 'Entre opposition et participation, les syndicats face aux réformes en Afrique'. *Revue Tiers Monde*, 224 (4), 9–24.

Ruteere, M. and Pommerolle, M.-E. (2003). 'Democratizing security or decentralizing repression? The ambiguities of community policing in Kenya'. *African Affairs*, 102, 587–604.

Ryan, M. (1978). *The Acceptable Pressure Group: Inequality in the Penal Lobby. A Case Study of the Howard League and RAP*. Westmead, Saxon House.

Saada, E. (2007). *Les Enfants de la colonie. Les métis de l'Empire français entre sujétion et citoyenneté*. Paris, La Découverte.

Salau, T. (2015). 'Public transportation in metropolitan Lagos, Nigeria: Analysis of public transport users' socioeconomic characteristics'. *Urban, Planning and Transport Research*, 3 (1), 132–139.

Salo, E. (n.d.). 'Ganging practices in Manenberg, South African and the ideologies of Masculinity, gender and generational relations'. unpublished art. University of Cape Town.

Salvaire, C. (2019). 'From urban congestion to political confinement: Collecting waste, channeling politics in Lagos'. *Territory, Politics, Governance*, online access.

Samara, T.R. (2010). 'Policing development: Urban renewal as neo-liberal security strategy'. *Urban Studies*, 47 (1), 197–214.

Sapire, H. (1992). 'Politics and protest in shack settlements of the Pretoria-Witwatersrand-Vereeniging region, South Africa'. *Journal of Southern African Studies*, 18 (3), 670–697.

Sapire, H. (1994). 'Apartheid's testing ground: Urban native policy and African politics in Brakpan, South Africa, 1943–1948'. *Journal of African History*, 35, 99–123.

Schack, W. and Skinner, E. (eds.) (1978). *Strangers in African Societies*. Berkeley (CA), University of California Press.

Seekings, J. and Lee, R. (2002). 'Vigilantism and popular justice after Apartheid'. In: *Informal Criminal Justice* (ed. D. Feenan). Aldershot, Ashgate.

Seekings, J. (1993). *Heroes or Villains? Youth Politics in the 1980s*. Johannesburg, Ravan Press.

Seekings, J. (2000). 'Urban studies in South Africa after Apartheid'. *International Journal of Urban and Regional Research*, 24 (4), 832–840.

Seekings, J. (2000a). 'Social ordering and control in the African Townships of South Africa: An historical overview of Extra-State initiatives from the 1940s to the 1990s'. In: *The Other Law: Non-State Ordering in South Africa* (eds. W. Scharf and D. Nina). Le Cap, Juta and Co.

Seekings, J. (2000b). *The UDF: A History of the United Democratic Front in South Africa, 1983–1991*. Claremont, David Philip/Oxford, James Currey.

Seekings, J. and Natrass, N. (2006). *Class, Race and Inequality in South Africa*. Scottsville, University of KwaZulu Natal Press.

Seekings, J. (2008). 'The Carnegie Commission and the backlash against welfare state-building in South Africa, 1931–1937'. *Journal of Southern African Studies*, 34 (3), 515–537.

Sen, S. (2004). 'A separate punishment: Juveniles offenders in Colonial India'. *The Journal of Asian Studies*, 63 (1), 81–104.

Serre, D. (2009). *Les Coulisses de l'État social. Enquête sur les signalements d'enfant en danger*. Paris, Raisons D'agir.

Sesay, A., Ukeje, C., Aina, O., and Odebiyi, A. (eds.) (2003). *Ethnic Militias and the Future of Democracy in Nigeria*. Ile-Ife. Obafemi Awolowo University Press.

Sharma, A. and Gupta, A. 'Rethinking theories of the state in an age of globalization'. In: *The Anthropology of the State. A Reader* (eds. A. Sharma and A. Gupta), 1–41. Oxford, Blackwell, 2006.

Shaw, C. (1929). *Delinquency Areas*. Chicago (IL), University of Chicago Press.

Shaw, C. and Mckay, H. (1942). *Juvenile Delinquency and Urban Areas*, Chicago (IL), University of Chicago Press.

Shear, K. (2012). 'Tested loyalties: Police and politics in South Africa, 1939–1963'. *Journal of African History*, 53 (2), 173–193.

Simone, A.M. (2001). 'On the Worlding of African cities'. *African Studies Review*, 44 (2), 15–41.

Simone, A.M. (2004). *For the City Yet to Come. Changing African Life in Four Cities*. Durham (NC), Duke University press.

Simone, A.M. (2010). *City Life from Jakarta to Dakar: Movements at Crossroads*. Durham (NC), Duke University Press.

Simone, A.M. and Pieterse, E. (2017). *New Urban Worlds, Inhabiting Dissonant Times*. Cambridge, Polity Press.

Sitas, A. (1996). 'The new tribalism: Hostels and violence'. *Journal of Southern African Studies*, 22 (2), 235–248.

Sklar, R. (1963, 2004). *Nigerian Political Parties. Power in an Emergent African Nation*. Trenton, African World Press.

Smith, N.R. (2019). *Resisting Rights. Vigilantism and the Contradictions of Democratic State Formation in Post-Apartheid South Africa*. Oxford University Press, New York.

Solomon, B. (2018). 'Occupancy urbanism: Radicalizing politics and economy beyond policy and programs'. *International Journal of Urban and Regional Research*, 32 (3), 719–729.

South African Institute of Race Relations (SAIRR) (1938). *Memorandum on Native Juvenile Delinquency*. Johannesburg.

South African Republic (1963). *Statistics of Offences and of Penal Institutions, 1949–1962*. Pretoria, Government Printer.

Spear, T. (2003). 'Neo-traditionalism and the limits of invention in British Colonial Africa'. *Journal of African History*, 44 (1), 3–27.

Stacey, P. (2019). *State of Slums: Precarity and Informal Governance at the Margins of Accra*. London, Zed Books.

Steck, J.-F. et al. (2013). 'Informality, public space and urban governance: An approach through street trading. Evidences from Abidjan, Cape Town, Johannesburg, Lomé and Nairobi'. In: *Governing Cities in Africa, Politics and Policies* (eds. S. Bekker and L. Fourchard). Pretoria, HSRC Press.

Steinberg, J. (2004). *The Number, One Man's Search for Identify in Cape Underworld and Prison Gangs*. Le Cap, Jonathan Ball.

Steinberg, J. (2008). *Thin Blue. The Unwritten Rules of Policing in South Africa*, 49–57. Le Cap, Jonathan Ball.

Suberu, R. (2001). *Federalism and Ethnic Conflict in Nigeria*. Washington (DC), United States Institute of Peace Press.

Suberu, R. (2013). 'Prebendal politics and federal governance in Nigeria'. In *Democracy and Prebendalism in Nigeria. Critical Interpretations* (eds. W. Adebanwi and E. Obadare). Basingstoke, Palgrave Macmillan.

Super, G. (2016). 'Volatile sovereignty: Governing crime through the community in Khayelitsha'. *Law and Society Review*, 50 (2), 450–483.

Super, G. (2020). '"Three warnings and you're out": Banishment and precarious penalty in South Africa's informal settlements'. *Punishment and Society*, 22 (1), 48–69.

Swanson, M. (1977). 'The sanitation syndrome: Bubonic plague and urban native policy in the Cape Colony, 1900–1909'. *Journal of African History*, 19, 387–410.

Taliani, S. (2012). 'Coercion, fetishes and suffering in the daily lives of young Nigerian women in Italy'. *Africa*, 82 (4), 579–608.

Tamuno, T.N. (1989). 'Trends in policy: The police and prisons'. In: *Nigeria since Independence. The First 25 Years* (eds. T.N. Tamuno and J.A. Atanda), 4. Ibadan, Heinemann.

Thenault, S. (2012). *Violence ordinaire dans l'Algérie coloniale. Camps, internements, assignations à résidence*. Paris, Odile Jacob.

Theroux, L. (2011) 'Law and Order in Lagos'. BBC documentary movie.

Thomas, M. (2007). *Empires of Intelligence. Security Services and Colonial Disorder after 1914*. Berkeley (CA), University of California Press.

Thomas, L. (2013). 'Violence and colonial order: Patterns of policing during the depression years', Policing Empires: Social Control, Political Transition, (Post)Colonial Legacies, Conference Royal Academy of Belgium, 12–13 December.

Truth and Reconciliation Commission of South Africa Report, vol. 3, 1998.

Tibenderana, P.K. (1983). 'The emirs and the spread of western education in Northern Nigeria'. *Journal of African History*, 24 (4), 517–534.

Titeca, K. (2014). 'The commercialisation of Uganda's 2011 election in the urban informal economy: Money, Boda-Bodas and Market vendors'. In: *Elections in a Hybrid Regime: Revisiting the 2011 Ugandan Polls* (eds. S. Perrot, S. Makara, J. Lafargue and M.A. Fouéré). Kampala, Fountain Publishers.

Tolimson, R. et al. (eds.) (2003). *Emerging Johannesburg: Perspectives on the Postapartheid City*. London, Taylor and Francis.

Ubah, C.N. (1985). *Government and Administration of Kano Emirate, 1900–1930*. Nsukka, University of Nigeria Press.

Ukiwo, U. (2006). 'Creation of local government areas and ethnic conflicts in Nigeria: The case of Warri, Delta State'. *CRISe Working Paper*.

Utas, M. (ed.) (2012). *African Conflicts and Informal Power: Big Men and Networks*. London, Zed Books.

Van De Walle, N. (2007). 'Meet the new boss. Same as the old boss?'. In: *Patrons, Clients and Policies: Patterns of Democratic Accountability and Political Competition* (eds. H. Kitschelt and S.I. Wilkinson). Cambridge, Cambridge University Press.

Van De Walle, N. (2014). 'The democratization of clientelism and social policy'. In: *Clientelism, Social Policy, and the Quality of Democracy* (eds. D. Abente Brun and L. Diamond). Baltimore (MD), Johns Hopkins University Press.

Van der Spuy, E., Schârf, W. and Lever, J. (2000). 'The politics of youth crime and justice in South Africa'. In: *The Blackwell Companion to Criminology* (eds. C. Summer). Oxford, Blackwell.

Van Onselen, C. (1979). *Chibaro: African Mine Labour in Southern Rhodesia, 1900–1933*. Johannesburg, Ravan Press.

Van Onselen, C. (1984). *The Small Matter of a Horse: The Life of 'Nogolozza' Mathebula, 1867–1948*. Johannesburg, Ravan Press.

Van Onselen, C. (2001) [1982]. *New Babylon, New Nineveh: Everyday Life on the Witwatersrand, 1886–1914*. Johannesburg, Jonathan Bell.

Varese, F. (2014). 'Protection and extortion'. In: *The Oxford Handbook of Organised Crime*. (ed. L. Paoli). Oxford, Oxford University Press.

Vaughan, M. (1991). *Curing their Ills: Colonial Power and African Illness*. Redwood City (CA), Stanford University Press.

Veit, A., Barolsky, V. and Pillay, S. (2011). 'Violence and violence research in Africa South of the Sahara'. *International Journal of Conflict and Violence*, 5 (1), 13–31.

Venter, J. (1959). *The Extent of Juvenile Crime in South Africa: An Analysis of the Problem Based on Available Information*. Pretoria, Department of Education and Social Research.

Venter, H.J. and Retief, G.M. (1960). Bantoe-Jeudmisdaad : 'n krimineel-sosiologiese ondersoek van 'n groep naturellejeugoortreders in die boksburgese landdrosdistrik. Department of Sociology, University of Pretoria.

Veyne, P. (2008). *Foucault, sa pensée, sa personne*. Paris Albin Michel.

Viljoen Commission (1976). *Commission of Enquiry into the Penal System of the Republic of South Africa*. Pretoria, Republic of South Africa.

Wacquant, L. (2004). *Punir les pauvres. Le nouveau gouvernement de l'insécurité sociale.* Marseille, Agone.

Waller, R. (2006). 'Rebellious youth in Colonial Africa'. *Journal of African History*, 47 (1), 77–92.

Ward, K. (2006). 'Policies in motion'. Urban management and state restructuring: The trans-local expansion of business improvement districts. *International Journal of Urban and Regional Research*, 30 (1), 54–75.

Watson, R. (2000). 'Murder and the political body in early colonial Ibadan'. *Africa*, 70 (1), 25–48.

Watson, R. (2003). *Civil Disorder is the Disease of Ibadan: Chieftaincy and Civic Culture in a Yoruba City.* Oxford, James Currey.

Watson, V. (2014). 'African urban fantasies: Dream or nightmares?' *Environment & Urbanization*, 26–1, 215–231.

Webert, C. and Murray, M.J. (2015). 'Building from scratch: New cities, privatized urbanism and the spatial restructuring of Johannesburg after apartheid'. *International Journal of Urban and Regional Research*, 39 (3), 471–494.

Weber, M. (1919). *Wissenschaft als Beruf, Politik als Beruf.* Hg. Wolfgang Schluchter.

Weber M. (1978). *Economy and Society. An Outline of Interpretative Sociology.* Berkeley, University of California Press.

Weinstein, L. (2014). "One-man handled': Fragmented power and political entrepreneurship in globalizing Mumbai'. *International Journal of Urban and Regional Research*, 38 (1), 14–35.

Wilcocks, R. W. (1932). 'The intelligence quotient of the poor white child', vol. 2. *Report of the Carnegie Commission: The Poor White Problem in South Africa.* Stellenbosch, Pro Eclesia Drukkery.

Wilson, M. and Mafeje, A. (1963). *Langa: A Study of Social Groups in an African Township.* Oxford, Oxford University Press.

Yahya, M. (2007). 'Polio vaccines: "No thank you!" Barriers to Polio eradication in Northern Nigeria'. *African Affairs*, 106 (423), 185–204.

Ya'u, Y.Z. (2000). 'The youth, economic crisis and identity transformation: The case of the yandaba in Kano'. In: *Identity Transformation and Identity Politics under Structural Adjustment in Nigeria* (ed. A. Jega), 161–180. Uppsala, Nordiska Afrikainstitutet and the Centre for Research and Documentation, Kano.

Yonucu, D. (2018). 'Urban vigilantism: A study of anti-terror law, politics and policing in Istanbul'. *International Journal of Urban and Regional Research*, 42.3, 408–422.

Young, C. (2004). 'The end of the post-colonial state in Africa? Reflections on changing African political dynamics'. *African Affairs*, 103, 23–25.

Zerah, M.-H., (2008). 'Splintering urbanism in Mumbai: Contrasting trends in a multilayered society'. *Geoforum*, 39 (6), 1922–1932.

Appendix 1
Dictionary

Baale, bale, balogun, oba, olubadan

The *oba*, in Yorubaland, are the sovereign leaders of a city. Each *oba* has its own title in each city: in Ibadan it is the Olubadan. In the nineteenth century and for much of the colonial period, under the authority of the Olubadan, a *Baale* was responsible for the administration of the city. He also acted as an intermediary between the *oba* and the lower-ranking chiefs who include warrior lineage chiefs *(balogun)* and neighbourhood chiefs *(bale)*.

Bachelor

The term refers to all unmarried male adults under customary law in South Africa. *Bachelor status* is often associated with the migrant condition housed in compounds and *barracks* even if in reality many *bachelors* could maintain one or more women.

Bantustan, reserves

As white South African settlers conquered more territories in the nineteenth century, smaller portions came to be exclusively reserved for African populations that became what was considered as native reserves in the twentieth century. From the 1950s onwards, these reserves were incorporated into the *Bantustans*, autonomous governments not recognised by the international community but developed by the South African government to maintain and strengthen white supremacy in South Africa. These territories at the periphery of the white fertile rural areas were, as

Classify, Exclude, Police: Urban Lives in South Africa and Nigeria, English Language First Edition. Laurent Fourchard.

early as the 1930s, overcrowded and considered the main areas of rural ghettoisation of the country.

Coloured

A minority group throughout South Africa but a majority in the Western Cape province, the *coloured* are descendants of various unions between European settlers and women of Koisan, Bantu origin or from the Dutch Colony of Java. Although they spoke the same language as the first white settlers (Afrikaans rather than English) and had early contact with this minority, they were not accepted into colonial society but occupied an intermediate place in the South African racial hierarchy. They did not need to have *pass* and employment preferences were granted to them in the Western Cape Region from the mid-1950s onward. Classified as one of four racial groups by the apartheid government to maintain white rule and maintain racial divisions, the *coloured* were subjected under apartheid to severe discriminatory measures: prohibition of interracial marriage, segregation in public places, abolition of the right to vote and the removal of hundreds of thousands of them on the outskirts of white cities.

Compounds (South Africa)

The compound system was introduced into Kimberley diamond mines in 1885, before being extended to the Witwatersrand gold mines at the end of the nineteenth century, to house African migrant workers during their contract period in the mines. Housing conditions were deplorable due to the notorious overcrowding in all-male dormitories. The aim was to control – and keep at low cost – the migrant labour force, limit their independence and isolate them from the rest of the urban population.

Compounds (Nigeria) or *ile* (in Yoruba)

Most Yoruba cities were divided in the nineteenth century, and in much of the first half of the twentieth century, into neighbourhoods, under the authority of a chief, known as a *bale*. These historical quarters are made up of a compound known as *ile* that constitutes the basic residential unit of a lineage or an extended family. The *ile* continued to play an important role in the everyday working of the historical neighbourhoods of Lagos Island and Ibadan city centre.

Hostels or barracks

Sleeping quarters reserved for male migrants under contract who come to work temporarily in a mine, in town or on the farm of a white settler in South Africa. These dormitories were managed either by the private owners of the mine or of a farm or by the municipality. They operated on the same principle of control as the mining compound introduced at the end of the nineteenth century and survived until the 1990s despite the abolition of migration control in the mid-1980s.

Indigenes/non-indigenes

The *indigenes* of a place in Nigeria are Nigerian citizens who can connect their ethnic, genealogical or ancestral roots to a community of people from that place. Other citizens, no matter how much time they spent in this place, are described as *non-indigenes*. The word *indigene* has no link with the French colonial word *indigène* (which should be translated as native). I prefer to keep the term in italics rather than translate it as 'indigenous' or 'autochthonous'. The origin of the term probably dates from the 1970s and probably replaces the old colonial categories of *natives*.

Indirect rule

Indirect rule is a method of exercising colonial power in governing through *native authorities* vested with substantial powers of justice, police, administration and taxation under the control of the British authorities. Initiated in India, experimented in Uganda and then Nigeria at the beginning of the twentieth century, indirect rule became the official policy of the British Empire and extended during the interwar period to the other African colonies of the empire.

Natives/non-natives

The term *'native'* brings together the subjects of the British crown under the jurisdiction of the native authorities as part of the indirect rule. Non-natives, in colonial Nigeria, refer to populations that are not under the jurisdiction of native authorities but are under the authority of either the colonial administration or a customary non-native authority of a locality. *Non-natives* are most often grouped in the city in specific neighbourhoods called *sabon gari* or *sabo*. There is no equivalent in French West African colonies of the term *'non-native'* which reflects the specificity of British indirect rule in Nigeria, to classify and differentiate and rule accordingly different types of subjects.

Pass and Reference book

The pass is an identity document made mandatory for slaves in the Cape Colony from the eighteenth century onwards, for African mine workers in South Africa at the end of the nineteenth century before being extended to all-male African workers, including foreigners in 1911. The *pass* was intended to direct cheap labour where employers needed it. The *pass* was replaced from 1952 by the reference book, a biometric document made in the service of the *great apartheid* that systematically removed from the cities or the mines the African populations deemed useless for the apartheid economy and forcibly sent to the reserves or the *Bantustans*. From the mid-1950s onward, the *pass* was extended to African women and young males over 16 years. The *pass* system was abolished only in 1986. It has been estimated that between 15 and 20 million people were arrested, imprisoned and deported to South African reserves between the beginning of the past century and 1986.

Sabon gari and sabo

The term means new city in Hausa language. The first *sabon gari* was drawn in 1909 in Kano to accommodate Nigerian subjects who did not fall under the authority of the emir of Kano, mainly the Christian populations of the south of the country. This system was then extended to both northern and southern Nigeria, where these neighbourhoods were exclusively reserved for populations from other Nigerian regions. These migrant populations were assigned to these neighbourhoods on the basis of their ethnic origin throughout the colonial period. Residents of Sabo or *sabon gari* referred to as *non-natives* enjoyed different rights from the *natives* and were either under the authority of a colonial administrator (district officer) or under that of the local non-native customary authorities.

Squat

A term used to describe spaces invested by Africans in white rural land and informal or irregular dwellings in South African cities. Urban squats were regularly demolished under apartheid before being further tolerated in the mid-1980s.

Shebeen

Illegal drinking spot that developed throughout South Africa with the gradual extension of the prohibition of alcohol and the regulation of beer consumption for all Africans from the end of the nineteenth century.

Slum

The term became popular at the end of the nineteenth century in the colonies of the British Empire. It refers to different realities in colonial times (unsanitary and epidemic conditions, overpopulation) and today (high population density, land insecurity, precarious building materials, lack of access to drinking water).

Index

Classify, Exclude, Police: Urban Lives in South Africa and Nigeria, English Language
First Edition. Laurent Fourchard.
© 2021 John Wiley & Sons Ltd. Published 2021 by John Wiley & Sons Ltd.

CPSIA information can be obtained
at www.ICGtesting.com
Printed in the USA
FSHW021210220521
81588FS